SUPERENERGY

AND THE QUANTUM VORTEX

Other books by David Ash

The Tower of Truth (with Anna Mary Ash) CAMSPRESS, 1978

The Vortex: Key to Future Science, (with Peter Hewitt) Gateway, 1990

The New Science of the Spirit, The College of Psychic Studies, 1995

Activation for Ascension, Kima Global Publishers, 1995

The Power of Puja, Puja Power Publications, 2004

The Power of Physics, Puja Power Publications, 2005

The New Physics of Consciousness, Kima Global Publishers, 2007

The Role of Evil in Human Evolution, Kima Global Publishers, 2007

The Vortex Theory, Kima Global Publishers, 2015

Continuous Living, Kima Global Publishers, 2015

AWAKEN, Kima Global Publishers, 2018

Acknowledgements

I wish to thank Susan Saillard-Thompson for her editorial inputs and guidance in the production of this book and Matthew Newsome for his edits and contributions in Section II. I also wish to thank Annemarie Jenson of Mill House for her work and her enthusiasm to bring SUPERENERGY and the physics of the Quantum Vortex into publication.

SUPERENERGY

AND THE QUANTUM VORTEX

DAVID ASH

WITH MATHEW NEWSOME

AND SUSAN SAILLARD-THOMPSON

Print book revised edition

ISBN 978-87-94302-73-9

First published by Kima Global Publishing S.A.

Associates: Mathew Newsome, Susan Saillard-Thompson

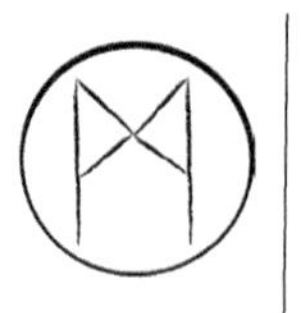

Published

Mill House Publishers, Denmark

https:www.millhouse-publishers.com

CONTENTS

FOREWORD

No doubt historians existing in the late 21st Century and beyond will look back to these years spanning the second millennium as a major turning point in the history of science and in human knowledge. We are currently living through this revolution and, without the benefit of hindsight, the picture can seem muddy as there are still many detractors alive today. But the wheels are in motion and the revolution is upon us.

I strongly believe that one of the key protagonists who will stand out when this rear-view mirror is used by future generations is David Ash. In the face of sometimes hostile ridicule from those who wish to perpetuate the dogma of Democritus, he has had the courage to repeatedly point out where, in the current paradigm, the emperor has no clothes. He has the insight to ask the questions that no one else dares or even articulates that maybe our current scientific models are based on flimsy non-fundamental principles, supported more by historic precedent and academic funding than nature.

So, amongst the crumbling of the old, there are some like Ash who are bringing in the new. His vortex theory has the insight to solve many of the issues in current science and he does this with rigorous logic making his ideas accessible to the scientist and non-scientist alike. Indeed, the theory has had the track record of predicting future scientific observations - a strong sign of its validity.

Ash disregards the prejudice of our current Western culture that sees ancient Eastern wisdom as somehow irrelevant and primitive and in doing so revives the true nature of wisdom: beyond the laboratory of modern science the universe is available for direct experience. In fact, he knows that although our experiments using the scientific method have indeed led us to great knowledge, the wisdom of ancient cultures can be deeper, as it is obtained from within.

You have an incredible adventure in front of you during which you can leave behind preconceived divisions between aspects of knowledge. Ash will take you far and wide on this epic journey – from deep into the atom, to meditating in India, to joining Alice in Wonderland.

Crucially, he is one of the guides of this new revolution, shaking us up to look deeper and realise that the universe is as it is: it is we who have compartmentalised knowledge and created dogmatic rules. So let him take you back to the truth.

Let the revolution begin!

- Dr Manjir Samanta-Laughton.

We have been duped

We have been duped. We have been duped by the cult of materialism – or physicalism as it is now called. Materialism has been disproved by science and materialistic values are destroying our spirit and our planet. We have been duped into believing that only what we can see and touch is real when quantum theory suggests the opposite is true. What we see is unreal. What we don't see is real.

Countess Amanda Fielding said, *"Science is the religion of the modern age; it killed off spirituality."*[1] It is not science, it is the cult of materialism masquerading as science that is killing off spirituality. The new science of superenergy based on the promising physics of the quantum vortex is helping to restore spirituality because spiritual beliefs and values are the soul of humanity; they are desperately needed in the world today.

- **David Ash** *May 2nd 2023*

1 The Financial Times HTSI, p.54, 29th April 2023

Out of the mouths of babes

Four year old Otto was asked by his father if he believed in God. He beat his little chest and replied *"God is in here and if we believe in heaven we go there, if we don't, we won't."*

When I was four, in the company of my mother, a lady bent down to me and asked what I was going to do when I grew up. Without hesitation I replied, *"I am going to prove the existence of God through science."*

BOOK I
SUPERENERGY

Put all your beliefs into harmony with science; there can be no opposition, for truth is one. When religion, shorn of its superstitions, traditions and unintelligent dogmas, shows its conformity with science, then there will be a great unifying cleansing force in the world which will sweep before it all wars, disagreements, discords and struggles – and then will mankind be united in the power of the Love of God.

'Abdu'l-bahá

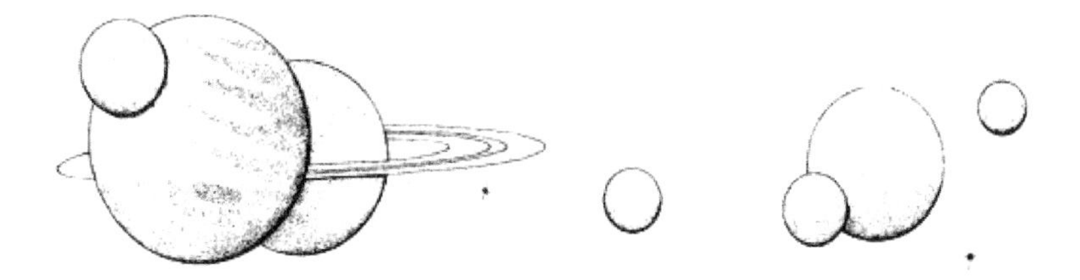

INTRODUCTION

Many people are under the impression that to believe in the supernormal is to be unscientific. That is not so. Scientific evidence is accumulating for the existence of other worlds beyond the physical. It is coming mostly from critical care medicine which reveals we do survive death. We live on and go elsewhere after our bodies die and a new science of the supernatural has emerged that can explain how this happens, where we go and a whole lot more.

On the 18th February 2022 the New York Academy of Science published a paper presenting evidence that Near Death Experiences (NDEs) are real and not hallucinations of dying brains.[1] An international team of scientists concluded that contrary to normal scientific belief, part of us survives death and goes somewhere else.

A panel of experts from the UK and USA, including critical care specialists, neuroscientists and psychiatrists from Kings College London, Baylor University Texas, Southampton University, New York University, University of California, University of Virginia, and Harvard University, put their names to the paper giving their professional opinion that people who have had an NDE actually survived clinical death. From a large number of case studies a clear pattern emerged. A percentage of people who revived from an episode of clinical death reported similar experiences, with common themes that are nothing like hallucination or illusion.

Many NDE reports follow a similar narrative arc, which includes separation from the physical body, after which, people find themselves still in a body but with a heightened sense of conscious awareness and perception. After looking down on their own lifeless body, from the body with heightened faculties, they travel to a destination which is either a place of darkness or a place of light. The majority go into a place of light where they report having a life review and often a feeling of coming home.

The evidence of near-death experiences shows that psychological and cognitive processes do not end with clinical death. The international team have concluded that the evidence, collected over many years from their clinical studies of NDEs, points to the mind being a universal reality that is not limited to the brain and that there is a continuity of life after death. We are faced with a stark choice. We either ignore or accept the evidence of NDEs.

---

1 Parnia S. et al., Guidelines and standards for the study of death and recalled experiences of death - a multidisciplinary consensus statement and proposed future directions. Ann.NYAS, Vol 1511, Issue 1, P.5-21, 18-02-2022

If, in the name of science we ignore evidence, then science could be accused of ignorance - taken that the word 'ignorance' is derived from the verb 'to ignore'. Alternatively, we can accept the evidence and question the philosophy of materialism that underlies science.

The consensus opinion in science is based on the philosophy of materialism, which is opposed to belief in the continuity of life after death. No matter how much evidence is presented, most scientists today will not shift from their opinion that near-death experiences are hallucinations of dying brains and they will never accept evidence to the contrary but a revolution is upon us which will put a fresh broom to science that will sweep away materialism.

In my teens I found a book on Yogi philosophy in an antiquarian collection that had been put to print before Einstein published his ground breaking equation $E=mc^2$. Thousands of years before Einstein, yogis discovered that matter is a form of energy. Peering into the atom with supernormal powers called *siddhi* they saw that the smallest particles in matter are whirlpools of light. That discovery led me to develop a new approach to quantum physics based on the yogic perception that subatomic particles are vortices of energy. The vortex physics revealed precisely how energy forms mass and showed that materialism is an illusion set up by spin.

As I applied my burgeoning vortex theory to subatomic matter, I realised it showed a way that supernatural matter could exist alongside physical matter. A hypothesis of super-speed energy unfolded alongside my quantum vortex theory which enabled me to explain supernormal phenomena.

Scientific materialism has extinguished hope of an eternal hereafter and brought in its place a morass of materialistic values that offer little more than sinking hearts and rising costs. We need hope and hope came to me when I realised that while subatomic particles could not move faster than the energy that forms them, if they were vortices of energy, the energy spinning in the vortices could exist with speeds in excess of the speed of light. I called the super-speed energy *superenergy.* This led me to predict the existence of worlds of superenergy of which the world of physical energy would be but a small part. That enabled me to make sense in science of near-death experiences, reincarnation and the supernatural.

CHAPTER 1
THE VORTEX OF ENERGY

My father was a physician and physicist. He introduced me to radioactivity physics when I was a boy. I have a vivid recall of an expedition to Bodmin moor in Cornwall when I was seven. He bundled me and my brothers, a sister and a cousin into his old Rolls Royce and took us prospecting for Uranium. He engendered in me a love of physics.

I was sixteen when I came across the *Advanced Course in Yogi Philosophy*[1] by Yogi Ramacharaka in an antiquarian library. It set me on a journey of discovery that would last a lifetime. I knew enough about physics by then to realise I had stumbled on something of immense importance when I read that yogis perceived atoms as vortices of prana. Knowing that prana is the Indian word for energy and realising the book was put to print before 1905, when Albert Einstein published his famous equation $E=mc^2$, I was galvanized.

It struck me immediately that if yogis could anticipate Einstein in realising the smallest particles of matter are forms of energy then in the vortex they may have uncovered the greatest enigma in science, how energy forms mass. As I followed the vortex thread through subsequent decades I realised yogis had discovered the key to the Universe.

Reading the 1904 book in 1964 I knew atoms weren't the smallest particles in matter. It had to be subatomic particles that were vortices of energy. I was determined to see if the protons, neutrons and orbiting electrons in the atom could be explained by a vortex theory.

I visualised subatomic particles in matter as whirlpools of light and as years of enquiry unfolded it became clear to me that the vortex was a store of energy. That enabled me to explain how vast amounts of nuclear energy could be released from minute amounts of matter. The gyroscopic spin of energy

1 Yogi Ramacharaka, An Advanced Course in Yogi Philosophy and Oriental Occultism, The Yogi Publication Society, 1904

accounted for the inertia of mass and interactions between extending dynamic vortices of energy explained the forces of gravity, magnetism and electric charge. The vortex is a three-dimensional spiral. I realised that is why we live in a three dimensional world. My head was forever spinning.

The quantum vortex, as I now call it, shows how non-substantial energy can set up the properties of materiality. In physics the stuff of material is defined in terms of mass, inertia, potential energy and three-dimensional extension. These properties of subatomic particles, generally attributed to material substance, can be accounted for by spin in the vortex of energy. The quantum vortex explains away materialism. The vortex of energy exposes materialism as spin. I could call my book, *The Material Delusion.*

Materialism is based on our perception that something or someone exists and then acts. Be it an atom moving or a god creating, the idea that something substantial preexists animation is a fundamental human belief. It is hard to grasp the idea that activity can exist without any thing preexisting it that acts. How can there be movement when there is no thing that moves? This conundrum may baffle lesser minds but it is not a problem to quantum physicists and other people who have grasped the principle that everything is energy. Enlightened people have dropped the mechanical view of the Universe and appreciate the non-substantial nature of matter. Albert Einstein led the quantum revolution that banished scientific materialism but he was not properly understood because, as William Berkson wrote, *Einstein was difficult to understand... not because of his ideas or the mathematics he employed, but because of his world view. Einstein denied the substantiality of matter and the field, whilst maintaining their reality.* [2]

Energy is no thing. As activity where nothing exists that acts; as movement without anything that moves, energy is an enigma. Richard Feynman said, *"It is important to understand that in physics today we have no idea what energy is."*[3]

Some physicists have overcome the enigma of energy by accepting that it is abstract in nature. A number of quantum physicists have realised that the unsubstantial particles of energy, underlying matter and light, are more like thoughts than things. At the height of the quantum revolution the English astrophysicist, Sir James Jeans, wrote: *Today there is a wide measure of agreement that the stream of knowledge is heading toward a non-mechanical reality; the universe begins to look more like a great thought than a great machine. Mind*

2 Berkson W. Fields of Force: The Development of a world view from Faraday to Einstein, John Wiley & Sons, 1974

3 Feynman R. Feynman Lectures on Physics Basic Books 2010

no longer appears as an accidental intruder into the realm of matter; we are beginning to suspect that we ought rather to hail it as the creator and governor of the realm of matter. [4]

The underlying nature of reality as non-substantial and dynamic was already appreciated in the East. Fritjof Capra summed this up in *The Tao of Physics* [5] and *The Turning Point*[6] when he wrote about Indian thinking that anticipated quantum theory: *There is motion but there are, ultimately, no moving objects; there is activity but there are no actors; there are no dancers, there is only the dance...The Vedic seers...saw the world in terms of flow and change, and thus gave the idea of a cosmic order an essentially dynamic connotation... Shiva, the cosmic dancer, is perhaps the most perfect personification of the dynamic Universe... The general picture emerging from Hinduism is one of an organic, growing and rhythmical moving cosmos; of a Universe in which everything is fluid and ever changing, all static forms being maya, that is, existing only as illusory concepts.*

Yogis became aware of the dynamic underpin of matter when they perceived the smallest particles to be vortices of energy. It is thanks to them we can appreciate how energy, that has no mass or material substance, can create the massive and seemingly substantial world we live in. The quantum vortex was seen by yogis through the operation of extranormal powers. These enabled them to perceive energy spinning to set up a static state. This led them to declare that everything material is in essence a condensed form of mind[7] and that manifest things are maya - the illusion of forms.

If we go beyond maya in our physical perception of reality and accept that the universe is a mind and particles of energy are more like thoughts than things then, because thoughts are particles of intelligence, we could conclude that particles of energy are particles of intelligence. This would imply that intelligence is innate in the quantum, which would in turn suggest that intelligence is universal. If the Universe were intelligent then it would seem to be obvious that universal intelligence could underlie the evolution of life. If that were so we might have to reconsider the popular presumption that the peak of intelligence is in the human head.

Quantum intelligence would predicate for intelligence at every level in the Universe. It would enable intelligence to be presumed in evolution without

4 Jeans J. The Mysterious Universe, Cambridge Uni. Press, 1930

5 Capra F., The Tao of Physics, Wildwood House, 1975

6 Capra F., The Turning Point, Fontana, 1983

7 Yogi Ramacharaka, Fourteen Lessons in Yogi Philosophy, Fowler

needing to assume the existence of a creative intelligence acting from outside of the Universe. Instead, we could imagine we are in a Universe that is evolving from the innate intelligence in its quantum fabric. This brave idea would allow for the possibility it is not just humans but the entire Universe that is driving forward systems of creative evolution capable of discovery, art and innovation. Surely this is evident. Is it not obvious that intelligence is innate in the myriad forms of life all around us.

There could be intelligence in the shape of the quantum of energy and memory in the multitude of forms resultant from quantum interactions. If the vortex and vibrational forms of energy are more in the nature of thoughts than things, intent could underly the process of evolution and the Universe might not be as pointless as many educated people presume. If the Universe is a collective of quantum drops of intelligence, because intelligence implies consciousness the Universe could be underpinned by an ocean of creative consciousness with awareness, intent and purpose in its depths. Maybe molecules in the matrix of living matter are forms of intended creative imagination. If that were so traditional ideas of spirit and creation could be reviewed in terms of creative potential and intent emerging from universal consciousness and quantum intelligence. Evolution could be seen as a process of learning through chance, trial and experiment by universal consciousness and intelligence. An appreciation of quantum intelligence could allow us to uphold consciousness as a universal principle without having to abandon science. From that standpoint I believe quantum physics would enable us to embrace the essential elements of spirituality in science without religious overtones.

Physicists believe the Universe is intelligent[8] and self-aware[9] and that consciousness underpins the world we live in. In ancient Greece Plato taught that consciousness created the world.

A generation before Plato, Democritus was teaching that matter was composed of indestructible material atoms. His insight into the particulate nature of matter was brilliant but his presumption that the randomly interacting atoms were substantial was wrong. Einstein challenged that presumption when he established that non-substantial particles of energy underlie atoms. Einstein established the particulate nature of energy. That is quantum reality. In his equation $E=mc^2$ he revealed that non-material particles of energy underlie

8 Hoyle F. The Intelligent Universe, Michael Joseph, 1983

9 Goswami A The Self-Aware Universe: How Consciousness Creates the Material World, Tarcher, 1993

matter as well as light. On one hand Einstein upheld the particle concept of Democritus, on the other he repudiated the idea of particle materialism that emerged in his philosophy.

Particle materialism is unscientific. To see why imagine rain falling to form streams tumbling down mountain sides in the Alps. These join the torrent of rivers that pass through hydroelectric plants where the fall of the water is converted into the spin of turbines and then the flow of electricity. The electricity is then fed into CERN where it is used to accelerate protons. As the protons collide, in the intersecting rings of the Large Hadron particle accelerator, their arrested motion is transformed into mass. The kinetic energy derived from the motion of falling water has been converted into new particles of matter. In the fall of rain and the tumble of streams, the torrent of rivers and the spin of turbines, the flow of electricity and acceleration of protons no material substance was transferred to form the new particles. Activity alone was transferred between each step of the process. As nothing went into the newly formed particles of matter but activity, obviously, they can be nothing but forms of activity; the activity we call energy. The vortex of energy reveals in an obvious way how activity can form particles of matter; how energy forms mass.

People have been told that matter is a form of energy but while physics has provided equations and the evidence for this, up until now there has been no simple visual model in physics to match the naïve realism of solid material atoms attributed to Democritus by the Roman Lucretius in his poem *De Rerum Natura*. The classic model of material particles cannot explain how mass is formed of energy whereas the vortex of energy shows how light can form matter and how spin can set up the illusion of a material particle.

The idea of the vortex of energy originated in yogic philosophy and in that same philosophy yogis taught that light is a dense level of thought. They contended that there are finer levels of mind beyond the world of density we live in with a more subtle light than the light we perceive. The success of the vortex applied to quantum physics gave me overwhelming confidence in the philosophy of yoga but to adopt yogi thinking about the higher levels of mind I had to plunge through the light barrier with the premise there may be worlds of *superenergy* existing beyond the speed of light.

CHAPTER 2
SUPERENERGY

Spiritual and supernatural phenomena are implicit in yoga and can be explained in a scientific context if allowance is made in physics for energy existing beyond the speed of light. In his theory of relativity Einstein established the importance of the speed of light in our world. He said that mass, space and time are relative to the invariable speed of light, measured at 299,792,458 meters per second.

In his equation $E=mc^2$ Einstein revealed that mass is relative to the speed of light. This is easy to understand if we view subatomic particles as whirlpools of light rather than balls of material stuff. He then went onto declare that the speed of light is the sole universal constant. He may have been wrong in that assumption. It may be that the speed of light, measured by scientists, happens to be the intrinsic speed of energy that makes up our world in this part of the Universe. In other parts of the Universe there could be worlds made up of energy with different speeds. This is possible in the quantum vortex theory.

The quantum vortex theory is based on the assumption that: *Everything in our world is formed of particles of movement at the speed of light in the form of vortex or wave.* This is why the laws of physics are relative to the speed of light. If that is so there could be other worlds based on speeds of energy faster than the speed of light spinning in vortices or vibrating in waves and the laws of physics pertaining to these worlds would be relative to the higher speeds.

The idea that energy could exist with speeds faster than that of common light may appear to contradict Einstein's theory of relativity. Relativity theory makes it clear that as particles accelerate, they don't exceed the speed of light - represented by the symbol c - they just become more massive. This is borne out by experiment. When particles are accelerated in high energy laboratories, they never move faster than the speed of light, they simply become increasingly massive and their mass increases exponentially as their velocity approaches that of light.

However, this restraint is based on the materialist idea that speed is a property of a material entity. If speed is a property of the energy forming entities then theoretically it is possible that the energy spinning *within* a quantum vortex could posses a speed faster than the common speed of light.

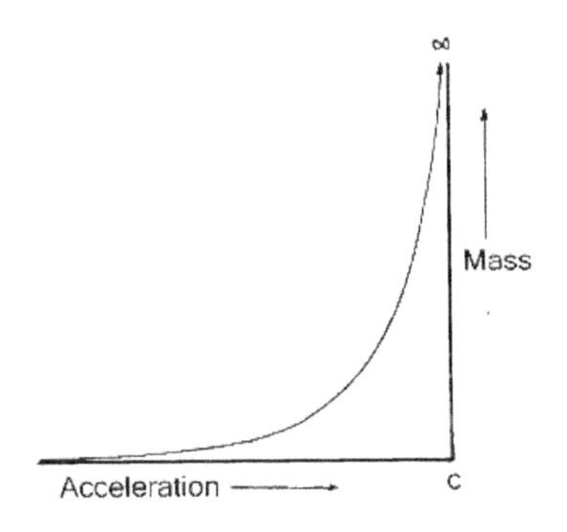

The superenergy hypothesis is based on the possibility that there could be worlds made up of energy spinning in vortices or vibrating in waves faster than the speed of light and there would be no reason to suppose the laws of physics in these worlds would be different to the laws of physics that underpin our world. If that were so the only difference between these worlds and our own would be the speed of the light whirling or undulating in the quantum vortices or waves that make them up. The speed of the energy underpinning each world would be its constant of relativity. I call these *The Einstein Constants* after Albert Einstein..

The superenergy hypothesis is an axiom that is not proved by scientific experiment. It is invented. This is fine because it is the start point of the superenergy theory and as Einstein said, "*The axioms of a scientific theory cannot be drawn from scientific experiment, they must be freely invented.*"

Superenergy may be just a theory but theories in science are useful if they help us make sense of evidence. The material hypothesis fails in the face of NDEs but the theory of superenergy can help us build a theoretical framework to account for life-after-death from the evidence of NDEs.[1]

To be acceptable in science a novel thesis must be reasonable and make sense of evidence that cannot otherwise be understood. The postulate that superenergy exists in vortices and waves to form worlds beyond the speed of light that are just like our own and are governed by the same laws of physics concurs with the NDE reports that after people die they find themselves in a world very similar to our own.

Many people imagine that if we go to another world after death it would be ethereal. However, in near-death experiences people report that the other world they visit is as real and seemingly substantial as our own.

---

1 Parnia S. et al., Guidelines and standards for the study of death and recalled experiences of death - a multidisciplinary consensus statement and proposed future directions. Ann. NYAS, Vol 1511, Issue 1, P.5-21,

Substantiality is a subjective experience. Our perception of substantiality is due in part to the inertia of mass, which is a consequence of spin in the quantum vortex of energy. It is also the way we experience atomic and molecular bonding. Bonding forces in atoms and molecules are properties of the vortex of energy which are explained in *The Quantum Vortex* in terms of vortex interactions. Substantiality could be perceived in the superenergy worlds if superenergy spins to form subatomic particles of matter and vibrates in waves to form photons of light. Spinning and vibrating superenergy could setup atoms, molecules and life forms in worlds full of light just like our own. Sentient beings could live in bodies like ours in these worlds and experience reality through them just as we do. Certainly this is what the NDE evidence suggests.

Thinking about superenergy in this way can help us to incorporate the supernatural in a scientific frame of under-standing. People embedded in a materialistic world view might not be excited by this but others who are more spiritually inclined may be enthralled by the arrival of a world view grounded in science that allows for higher dimensions of reality. Because of the bias toward materialism in science most information about higher dimensions comes from sources outside of science such as the *Urantia* book, which provides useful pointers in our line of enquiry.

According to the Urantia book[2] there are other worlds based on higher speeds of light and there is a world immediately beyond our own, of which ours is but a part. It is governed by an Einstein constant approximately twice the speed of our light. I call this world the *hyperphysical plane of reality*. Formed of *hyperphysical energy* it would correspond to the fourth dimension in common parlance. I believe it is the hyperphysical plane of reality that people visit when they have a near-death experience and where people go after physical death.

The hyperphysical plane associated with the Earth could correspond to the *anima mundi*. According to esoteric literature it seems that for the most part people do not leave the Earth after death but go to rest in a higher level of reality associated with the planet, which is known as the anima mundi. People reside there between lives and from there they may reincarnate for new life, in a fresh body, on the physical plane of the Earth which is formed of *physical energy*.

The idea that we incarnate from a hyperphysical plane of reality would make sense of the animating principle known to Plato as the *psyche*. Plato taught that a psyche, otherwise known as a *soul*, comes into incarnation at conception and departs at death. Plato's teaching of the psyche is supported by NDE evidence of human cognition surviving death of the physical body.

2 The Urantia Book, The Urantia Foundation (1999)

According to the Urantia book there are worlds beyond the hyperphysical level of reality which are governed by Einstein constants considerably higher than the speed of light. I call these levels of superenergy the *superphysical* planes of reality, the lowest of which corresponds to the fifth dimension in common parlance. In the Urantia book it says the fifth dimensional level is governed by an Einstein constant of approximately sixteen times the speed of light. According to the Urantia book there are planes of reality beyond the fifth dimension based on even higher speeds of energy. These higher planes of reality are not restricted to the Earth but are associated with deep space predominating the Universe at large.

Many people believe in spiritual dimensions, where souls go after exceptional incarnations on Earth. The science of superenergy supports the idea of spirits in higher dimensions of reality. These living forms of superphysical energy could correspond to the angels and gods spoken of in religion. Superphysical beings could also be the spirits that incarnate as human beings.

I believe the psyche or soul is a spirit that reincarnates many times leaving bodies of memory in the anima mundi of the Earth, somewhat like video data stored in the hard drive of a computer that can be accessed anytime. Some people describe these lifetime soul memories as *soul fragments.* I call them *soul bodies.*

I cannot prove the existence of the psyche, spirits or souls. I am not concerned to convince anyone of the existence of hyperphysical or superphysical dimensions. My endeavour is to make sense of traditional spiritual knowledge in terms of the superenergy model in the light of the accumulating evidence of near or after death experiences reincarnation and out of body experiences.

Many people who have had an NDE report the world they visited was occupied by deceased relatives and friends in recognisable bodies. Recognisable relatives could be accounted for by suggesting we grow a superenergy body in precise parallel to the physical body, which separates from the physical body when it dies much like a snake shedding its skin. The super-body, corresponding to the soul body or soul fragment, could then go off into the world to which it belongs.

Suggesting we grow a superenergy super-body in parallel to the physical body could also help explain the *two-body-dilemma* some people have, during a near-death or out-of-body experience, when they find they are still in a body after they have died. The superenergy model allows for the possibility that we could be living simultaneously in parallel bodies, in the physical and hyperphysical dimensions. This premise of parallel bodies can help people who have had an NDE to understand how they could be alive and still in a body while looking down on their lifeless body. This situation was described by an American professor of art, Howard Storm.

In his book, *My Descent into Death*,[3] Howard Storm tells how he found himself in a body, with heightened faculties of perception looking down on a body, identical to himself, lying on the hospital bed where he had been moments before. This was a very bewildering experience for him as it is for many other people having similar experiences.

The fact that Howard Storm's new body was identical to his dead physical body suggests his physical body may have acted as scaffolding for his superenergy body. In his NDE Storm found himself in hell. The description of his life before he died suggests his physical life may have determined not only the form of his super-body but also its fate. NDEs report dark experiences as well as light. Fortunately they are rare but they raise concerns as well as questions.

How is it possible for us to have two bodies existing simultaneously? How can they be exactly identical, atom for atom, molecule for molecule and cell for cell? Is one physical and the other hyperphysical? If so does the hyperphysical, super-body correspond to the Platonic psyche or soul that is believed to overlay the physical body during physical life and separate from it at death? Does it correspond to a ghost? Answers to some of these questions can be derived from a principle I call the *principle of coincidence.*

According to the quantum vortex theory, the quantum vortex of energy sets up the space and time on our plane of existence. If there are multiple planes, space-time could be set up by vortex motion on each plane. If space-time is set up by the quantum vortex on each plane it could not exist between the planes. If the physical, hyperphysical and superphysical planes of reality are not separated by space or time they could be separated by the intrinsic speeds of their energy; their Einstein Constants. If the energy planes are separated by speed and not by space-time they could coincide; that is they could exist simultaneously in the same 'here and now'. Sheets of paper impaled on a spike depict the coincident planes of reality.

Spiritual lore has it that: "*There are angels all around us but we are not aware of them because they are moving too fast and we don't see the fairies either because we are too slow and clumsy.*" People in times gone by believed we are surrounded by spiritual beings that are coincident with us. They live in a parallel world but we are not aware of them because we are separated from them by a factor of speed.

3 Storm H. My Descent into Death, Clairview, 2000

Some people catch a glimpse of a hyperphysical body as ghost. Children and sensitive people claim to see fairies or angels but most people don't see spiritual entities in higher dimensions. Why is that so? The answer could be in another principle called the principle of *energy-speed subsets.*

It is self-evident that slow speeds are a subset or a part of fast speeds, i.e., bicycle speed is a subset of the speeds available to a car. A car can move at bicycle speeds but no matter how hard you pedal on your bike you won't catch an accelerating car. The principle of energy speed subsets, based on the law that slow speeds are a subset or a part of faster speeds, suggests that worlds based on lower speeds of energy are part of worlds based on higher speeds. That infers a nested relationship between the planes of energy.

Like nested Russian dolls the physical world could be nested in the hyperphysical which in turn would be nested in the superphysical. This arrangement of superimposed worlds of energy, with each world based on a lower energy speed nested in the worlds based on higher speeds of energy, can also be depicted as concentric spheres.

The physical world, based on the speed of light, is depicted by the central

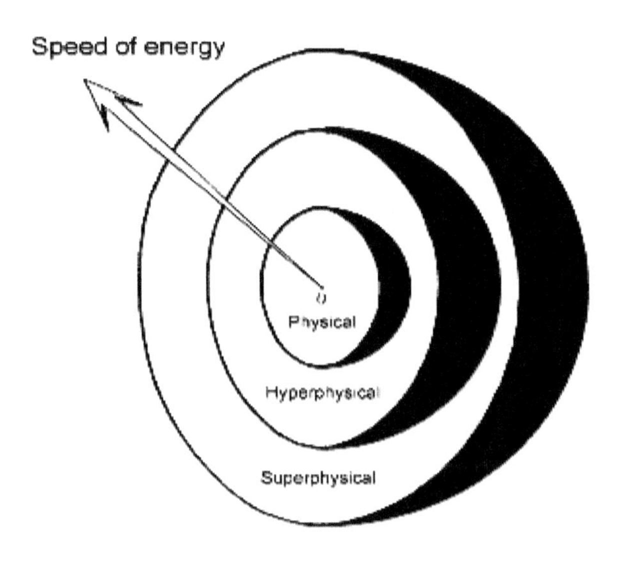

sphere. The hyperphysical world is represented by the second sphere and the superphysical corresponds to the third sphere out. This concentric sphere model of the physical, hyperphysical and superphysical levels of reality corresponds to the Pythagorean, *Harmony of Spheres.*

The harmony of the spheres depicts the planes of energy where the laws of physics are the same but only the intrinsic speeds of energy, the Einstein constants of relativity, differ. In *The Vortex: Key to Future Science*[4] my co-author, Peter Hewitt, proposed a matchbox analogy to illustrate the superimposed, nested relationship of the worlds.

He suggested we depict the physical world as a matchbox and imagine the matches in the match box as people that are only aware of the world inside the match box. They are not aware of the room outside the matchbox but the folk in the room are aware of them. In his analogy the room represents the hyperphysical level of reality, which in turn is part of a house, depicting the superphysical level of reality. You cannot fit a room into a matchbox and you cannot fit a house into a room. So it is, that movements with speeds beyond the speed of light cannot be contained within the physical space-time continuum, relative to the speed of light.

The perception of supernatural phenomena is limited by the principle of energy speed subsets. A supernatural or superenergy being could be in your presence but with your physical sight you would not see it. This is because physical energy does not interact with superenergy. Superenergy exists outside of physical space-time, so photons of physical light would not reflect off the superenergy being. Physical light would pass right through it, which is why it would be invisible to you. But it could see you because its superenergy light would reflect off you. This is because you and everything in your environment would be part of its space-time continuum.

These descriptions of coincident levels of non-substantial reality fit closely with the first nation Australian description of *dream states.* Living close to nature maybe they could see the nature of reality in a way that we in the civilized West cannot. The Western idea that *seeing-is-believing* may place a serious limitation on our perception of reality if our perception is limited by our civilized way of living. Things could be going on all around us which we cannot see because our senses have been dulled by the materialistic way of life.

4 Ash D. & Hewitt P. The Vortex: Key to Future Science, Gateway Books, 1990

People living close to nature may have more highly attuned senses required for survival in the wild. People who work on themselves or are naturally gifted to be more sensitive to coincident realities may also perceive energies and entities most of us are unaware of. The *flat-earth* mentality of diehard physicalists in materialistic civilisation can cause them distress when they die totally unprepared for their departure into another level of reality. They may find themselves standing in a hyperphysical body unable to communicate to a relative sitting by their empty physical shell that they are no longer in pain, sick or dying. Professor Howard Storm was a typical physicalist. He experienced this when he died.

In 1985, Howard Storm was in a hospital bed in Paris following a perforated duodenal ulcer. As described in, *My Descent into Death*, an operation should have been performed within five hours but it was a weekend and the hospital had only skeletal staff so he was left for ten hours before surgery. Howard was in terrible pain with only his wife Beverly by his bedside to comfort him. Suddenly he found himself pain-free standing by his bed. He said the floor was cool beneath his feet, the light was bright and everything in the room was crystal clear. He was wide awake and not dreaming. Hospital smells assailed his nose. All his senses were heightened and alert. He clenched his fists to prove he was awake and not dreaming. He was aware of blood coursing through his body and the beat of his heart was pounding in his ears. He was very anxious and his mind was racing but he felt more alive than he had ever felt before. He spoke to Beverly but she didn't react. He shouted at her but she still didn't respond, then he noticed a lifeless body on his bed under the sheet and was surprised at its resemblance to his own. It couldn't have been him because he was standing over it, looking down at it. Storm's initial reaction was annoyance that someone had played a nasty trick and placed a wax model of him in the bed where he had been lying, so ill and in so much pain, only moments before. He was very confused and leaning down to his wife he screamed at her until spittle was flying in her face but she continued to ignore him. She just sat in her chair weeping in despair.

CHAPTER 3
NEAR DEATH EXPERIENCES

Evidence of life after death is coming from rapid advances in resuscitation medicine, which have led to the ever-increasing number of near-death experiences. This term was first used by Elisabeth Kübler-Ross then by Dr. Raymond Moody, one of the first medical doctors to record extraordinary reports coming from patients who had been resuscitated.

In his landmark book *Life after Life*,[1] Raymond Moody presented numerous case study testimonials in support of the idea of life after death, collected from people who had a near-death experience. He described a typical near-death experience: *A man is dying and as he reaches the point of greatest physical distress, he hears himself pronounced dead by his doctor. He begins to hear an uncomfortable noise, a loud ringing or buzzing, and at the same time feels himself moving very rapidly through a long dark tunnel. After this, he suddenly finds himself outside his own physical body, but still in the immediate physical environment, and he sees his own body from a distance, as though he is a spectator.*[1]

Many people who have been through a near-death experience, in an operating theatre or a motor accident, report being out of their bodies, fully conscious, looking down on the scene of the operation or the accident: *I saw them resuscitating me. It was really strange. I wasn't very high; it was almost like I was on a pedestal, but not above them to any great extent, just maybe looking over them. I tried talking to them but nobody could hear me, nobody would listen to me...*[1]

Moody reported that many patients were aware of what people were saying and thinking about them: *I could see people all around, and I could understand what they were saying. I didn't hear them audibly, like I'm hearing you. It was more like knowing what they were thinking, but only in my mind not in their actual vocabulary. I would catch it the second before they opened their mouth to speak...* [1]

---

1 Moody R. Life after Life, Bantam Books, 1967

Near-death experiences support the idea that the mind exists as a universal reality and that we are capable of visiting other worlds. They suggest that when the brain dies, the cognitive mind is released from containment in the human body. Only by accepting mind as a universal principle can near-death and out of body experiences be understood. This flies in the face of scientific orthodoxy which insists that the mind is a product of molecular, neurological function and cannot survive the death of the brain. Eminent doctors like the neurologist, Sir John Eccles, disagree with the materialists, *"I maintain that the human mystery is incredibly demeaned by scientific reductionism, with its claim in promissory materialism to account eventually for all the spiritual world in terms of patterns of neuronal activity."*

Many doctors, scientists, and philosophers refuse to accept the accumulating evidence of near-death experiences, preferring to dismiss them as hallucinations of dying brains but experts in the field of resuscitation medicine say that is not possible if the patient is clinically dead when they have a near-death experience. Sam Parnia prefers to call it an *after death experience.*

Sam Parnia is a doctor whose specialty is resuscitation. In his book *Erasing Death,*[2] he explains that he doesn't start treating his patients until after they have been pronounced clinically dead. He contends that the term near-death is inaccurate because, from his professional point of view, people who have an NDE actually die and have an after-death experience, not a near-death experience.

Parnia said that half of the leading hospitals in Europe and America have resuscitation teams trained to bring people back to life after they have passed through clinical death. He said that worldwide the accumulated reports of an afterlife, from people being brought back from the dead, now number in millions. Dr Parnia and other critical care specialists routinely interview patients asking if they had an experience prior to resuscitation. Parnia said that about ten percent of people recall a vivid, life changing experience of another world beyond the one we live in. At the time of publishing, he had collected about five hundred case studies. The recalls often involved having a life review, traveling through a tunnel, experiencing a loving being of light and seeing relatives before being drawn back into the body.

Most people who have had an NDE report the other world is not dissimilar to our own. It had more light and beauty and felt like returning home. Only a few had an unpleasant experience of darkness and confusion. Many said people in the other world have bodies like our own. They saw or were greeted by relatives, often grandparents, who they recognised.

---

2 . Parnia S. Erasing Death, Harper One, 2013

In his book Dr Parnia claims that science is divided over the extraordinary experiences recalled by people who have actually died and been returned to life by the advances in resuscitation medicine. Parnia points out that attempts to explain away near-death experiences as hallucinations of a dying brain do not stand in cases where the person is clinically dead before resuscitation begins.

Sam Parnia is part of an international team of specialists dedicated to proper scientific approaches to NDEs. Recently, in a paper published in the Annals of the New York Academy of Science,[3] they explained that EEG brain scans note high gamma activity and electrical spikes in the brains of dying patients after death. This activity is quite the reverse of what is expected when a person is clinically dead. The researchers said that studies have been unable to disprove the experiences of people resuscitated, who have had a near-death experience.

In her PhD thesis on *The Near-death Experiences of Hospitalized Intensive Care Patients*,[4] based on a five-year clinical study, Penny Sartori explained away all the usual explain-aways. In her book, *The Wisdom of Near Death Experiences*, [5] she makes it clear that near-death experiences cannot be easily explained away if the subject is approached scientifically.

---

3 . Parnia S. et al., Guidelines and standards for the study of death and recalled experiences of death - a multidisciplinary consensus statement and proposed future directions. Ann.NYAS, Vol 1511, Issue 1, P.5-21, 18-02-2022

4 Sartori P. The Near-death Experiences of Hospitalized Intensive Care Patients: A five-year clinical study, Edwin Meller, 2008

5 Sartori P. The Near-death Experiences of Hospitalized Intensive Care Patients: A five-year clinical study, Edwin Meller, 2008

CHAPTER 4
REINCARNATION

Research into reincarnation has been carried out by a number of scientists, including Professor Ian Stevenson of the University of Virginia. Stevenson researched reincarnation extensively over a period of twenty years. He collected over 2,000 case studies in support of reincarnation and from 1960 onwards he published his findings in numerous journals and more than twenty books.[1]

Dr Stevenson concentrated his researches on the testimony of small children[2] and he took an interest in the relationship between birthmarks and past lives.[3] He chose mostly to study cases where children had spontaneous recollection of a past life that revealed details, which later could be cross referenced.

An outstanding case study involved a boy in France. He was born with a number of small birthmarks and as soon as he could speak he claimed the marks were left by bullets that killed him. As his speech developed he named the man who accused him of cheating at cards and then shot him. He detailed members of his family, the name of his girlfriend and the village where he lived in Sri Lanka. His French parents found he was a difficult child because he insisted on eating with his fingers and demanding rice and curry. He wrapped a cloth round himself fastening it like a sarong, broke into Singhalese and attempted to climb trees in search of coconuts. Subsequent investigation revealed that a coconut picker had been shot for cheating at cards, in the Sri Lankan village named by the boy, a few years before his birth in France. After the age of five the memories began to fade and the boy grew up as a normal French child.

Dr Karl Muller also investigated cases where children spontaneously recalled

---

1 Stevenson I. Twenty Cases Suggestive of Reincarnation, University of Virginia Press, 1988

2 Stevenson I. Children Who Remember Previous Lives: A Question of Reincarnation, McFarland & Company, 2000

3 Stevenson I. Reincarnation and Biology: A Contribution to the Etiology of Birthmarks and Birth Defects Volume 1: Birthmarks 1997

past lives.[4] He said it is not uncommon for children between the ages of two and four to communicate and behave as though they have had a previous existence. The problem with these cases is children of that age find it hard to express themselves and are rarely taken seriously by their parents and by the time they are articulate the memories tend to fade.

An Icelandic professor of psychology, Dr. Erlendur Haraldsson, started investigating reincarnation with Dr Ian Stevenson. He then went on to do his own ground-breaking research with children who not only recalled episodes from past lives but remembered coming into incarnation. Over twenty years he collected more than hundred case studies.[5]

In His book, *Reincarnation: Based on Facts*,[4] Dr Muller highlighted a graphic example of the religious atrocities that led to the rise of scientific materialism. In 1952 a man experienced agonizing stomach pains while watching a caged monkey in Zurich Zoo. After five months without relief his pains were diagnosed as psychosomatic and meditation was recommended. After a week of practice he had a vision which involved sight, smell and sound. He was aware of being tied to a stake in a medieval town. A dignitary from a church tribunal was reading an accusation from which he only recorded his name as Jan van Leyden. Then an executioner, dressed in a red cloak disembowelled him with red hot pincers. He lost consciousness and woke up to find himself in a mutilated state, suspended in a cage high above the town. In the cage he died an agonizing death. Two other victims that suffered the same fate died alongside him in separate cages. Four weeks after this dreadful vision the abdominal pains faded away.

The vision prompted him to undertake several weeks of research. Eventually in a library he uncovered the story of Jan van Laden executed in 1536 in precisely the barbaric manner he had relived in his vision. With the story was an illustration of three cages hanging from a church tower.

In his book *Lifetimes*,[6] Dr. Frederick Lenz said people who have past life recalls often identify the physical body as a prison that traps the 'real me' for the duration of life on earth. Innumerable people who have out-of-body, near-death or other-world experiences, identify more with being consciousness and mind than a physical body. As their consciousness breaks free from the body many of these people experience it as though waking from a dream: *I found myself in a vast place. I felt as though I had come home. I had no apprehensions, fears or*

4 Muller K., Reincarnation: Based on Facts, Spiritual Truth Press, 1970

5 Haraldsson E, Matlock J. I Saw A Light And Came Here: Children's Experiences of Reincarnation, White Crow 2017

6 Lenz F. Lifetimes, Bobbs-Merrill, 1979

worries. I no longer remembered my former life on earth. Nothing existed for me but quiet fulfillment. I was not conscious of time in the usual sense; everything seemed timeless. I felt as if I had always been there. It was similar to the feeling I have when I wake from a dream that seemed very real only to realise it wasn't real, it was only a dream. That is how I felt. My former life on earth had been a passing dream which I had now wakened from.[6]

According to Lenz, people who recall life beyond the physical body experience it as coming into a greater level of consciousness. They speak of a passage between levels akin to the levels of superenergy I have described. Lenz reports someone saying, *"I did not have the sense of moving through space. Everything was consciousness and pure awareness...I moved through thousands of levels. On each level different souls were resting before being born again. The lower levels were much darker. I somehow knew that the souls on these levels were not as mature as those on the higher levels. Finally I reached a level that I was comfortable on. I stayed there. I sensed there were many levels above the one I stopped at and that souls more advanced than I would go there."*[6]

Reports of consciousness surviving death are too numerous and universal to ignore. It makes sense that if consciousness can come into a physical body at conception and leave it at death it should be able to repeat the process many times to benefit from different experiences offered by various cultures in the procession of civilizations down through the ages. Just as a child cannot learn all its lessons in a single day at school, so it would be reasonable to suppose that an individualisation of consciousness may need to incarnate in a physical body a number of times, to gain full benefit of the lessons to be learnt from physical life on earth.

CHAPTER 5
CHALLENGING BELIEFS

In his book Sapiens: *A Brief History of Mankind,*[1] Yuval Noah Harari wrote about the importance of stories and myths in the success of humankind: *Any large-scale human cooperation – whether a modern state, a medieval church, an ancient city or an archaic tribe – is rooted in common myths that exist only in people's collective imaginations...Yet none of these things exists outside the stories that people invent and tell one another. There are no gods in the Universe, no nations, no money, no human rights, no laws and no justice outside the common imagination of human beings.*

If gods are extraterrestrials there would be gods in the Universe. If that were so Harari's dogmatic statement, that they are figments of our collective imagination, would be just his story. Harari didn't mention the stories in science. As a professor of history he should have an understanding of the history and philosophy of science. If he did he would have added something to the effect that, *there are no scientific theories outside the common imagination of human beings.*

Science is built on stories called axioms, hypotheses and theories. A major problem we have today is that many educated people, especially the more highly educated, don't seem to appreciate that scientific theories are stories. They seem to believe they are true. Harari fell into this trap. In his more recent book *21 Lessons for the 21st Century,*[2] while he draws on practically every sphere of human belief, especially religion, to reveal the extent to which we have been taken in by stories he failed to include the story of materialism. The reason is obvious. He believes in it. In *21 Lessons* Harari wrote: *In itself the Universe is only a meaningless hodgepodge of atoms.* That is a summary of the philosophy of materialism, or physicalism, which most scientists and academics believe.

1 . Harari Y.N. Sapiens: A Brief History of Humankind, Vintage, 2014

2 Harari Y.N. 21 Lessons for the 21st. Century, Jonathan Cape, 2018

Despite being overturned by Einstein and disproved in quantum physics, belief in the tenets of materialism increases as the number of people receiving higher education increases. It seems science is causing education to funnel people into the cult of Democritism, encouraging them to scorn anyone who question its dogmas. This is what religions do. It is a human trait. Sapiens are quick to question the beliefs of others but are not so quick to question their own beliefs.

Progress is made not by following mainstream beliefs but by challenging them; and that is why contesting theories and treating them as stories is so important in science. Albert Einstein didn't make his breakthroughs by building on the theories that were accepted in his day. He started a revolution in physics by overturning the tables in the temple of science. He came out with his own radical view of reality based on a momentary flash of genius then challenged the accepted axioms of science to make room for his own ideas. To quote Lincoln Steffans: *We know there is no absolute knowledge, that there are only theories; but we forget this. The better educated we are, the harder we believe in axioms. I asked Einstein in Berlin once how he, a trained, drilled, teaching scientist of the worst sort, a mathematician, physicist, astronomer, had been able to make his discoveries. "How did you ever do it" I exclaimed, and he, understanding and smiling, gave the answer: "By challenging an axiom."* [3]

In *The Quantum Vortex* I challenge the material axiom of Democritus that is endorsed by science, and replace it with a revolutionary idea that subatomic particles are not solid corpuscles of material substance but are ball vortices of energy that set up an illusion of substantiality. This new axiom has led me conclude that particles of energy are more like thoughts than things, intelligence is innate in the quantum and energy can exist beyond the speed of light. If particles of matter are vortices of physical energy it would make sense that they could not move faster that the physical speed that forms them. But there is no reason why energy faster than light, *within* the forms of vortex and wave, could not set up worlds beyond the speed of light. That could account for soul and spirit and life-after-death and show that these religious ideas may not be 'just stories'.

Most religions are based on belief in the existence of other worlds apart from the physical world we live in. Many also contend that a supreme being in a higher state of reality is responsible for the creation of the Universe and life within it. To counter this belief scientists have proposed a theory of evolution to account for life. There are elements of truth in both these hypotheses that can serve us but the theory of evolution in science and the story of creation

3 Steffans L., Autobiography, Harcourt Brace (N.Y.) 1931

in monotheistic religions are both flawed. The religious problem lies not so much in the story of creation because the sequence of events is correct and even the timing of creation in a week makes sense because in the theory of the quantum vortex time has been shown to be relative to size so billions of years to man could be as days on the scale of the Universe. The main problem in Western religions is monotheism.

CHAPTER 6
CREATION

Western monotheism is based on the idea that an intelligent designer created the Universe and life within it. This is rooted in the Bible, which begins with the verse: *In the beginning God created the heaven and the earth.* (Genesis 1:1)

In the next first twenty-four verses of Genesis, God is presented as the creator of all creatures, upon the Earth including humankind However, with the creation of man God is no longer presented as a single god but as the head of a group of gods: *Then God said, "Let us make man in our image, according to our likeness..."* (Genesis 1:26)

The Bible is inconsistent in regard to monotheism. Early in the book of Genesis, God appears more as a committee than a single individual. Something happened to the Bible in its history. Wherever the term *God* appears in modern Bibles the word *'Elohim'* occurs in ancient versions of the Old Testament. The term Elohim is dominant in the second oldest of the four main Pentateuch sources (five Books of Moses) generally known as the *Elohist Tradition* in which Elohim was used to refer to the Creator. In Hebrew the term *el* referred to an all-powerful God and *elohim*, being the plural of *el*, literally meant *the all-powerful gods.*

At some point in the history of the Bible a redaction occurred, which led to monotheism. One reason for this is there may have been a power struggle amongst the Elohim and one of them succeeded in taking control. A recent authoritative account of this redaction, which also points to the very real possibility of extraterrestrial involvement in the origins of humanity, is presented by Paul Wallis in his trilogy *Escaping from Eden,*[1] *The Scars of Eden*[2] and *Echoes of Eden.*[3] Whatever happened in the Bible it is clear that someone known as *the Lord* was in control. This lord amongst the Elohim is supposed to have taken responsibility for the creation of mankind and appeared to take charge of every situation, which included establishing a safe haven in which to contain his new creature: *The Lord God planted a garden eastward in Eden, and there he put the man he had formed.* (Genesis 2:8)

The leader of the Elohim laid down the law for mankind, commanding that humans should be kept in innocence and not have knowledge of good and evil: *And the Lord God commanded the man, saying; "From every tree of the garden you may freely eat; but from the tree of knowledge of good and evil, you shall not eat, for in the day that you eat the fruit of it you shall surely die."* (Genesis 2:16-17)

In Genesis 1:27 it stated that man and woman were created together, both male and female, in the image of the Elohim. But a possible insight into the nature of the Elohim can be gleaned from an interaction between the woman and a serpent:

Now the serpent was more cunning than any beast of the field which the Lord God had made.

And he said to the woman, Has God indeed said you shall not eat of every tree in the garden?

And the woman said to the serpent, We may eat the fruit of the trees in the garden: but of the tree which is in the midst of the garden, God has said, You shall not eat it, nor shall you touch it lest you die.

And the serpent said to the woman, You will not surely die: For God knows that in the day you eat of it, your eyes will be opened, and you will be like God,

1 Wallis P., Escaping from Eden, Axis Mundi Books, 2020

2 Wallis P., The Scars of Eden, Sixth Books, 2021

3 Wallis P., Echoes of Eden, Paul Wallis Books, 2022

knowing good and evil.

And when the woman saw that the tree was good for food, that it was pleasant to the eyes, and a tree desirable to make one wise, she took of the fruit thereof and did eat, and gave also to her husband with her and he did eat. Then the eyes of both of them were opened... (Genesis 3:1-7)

This is an extraordinary story. A serpent appeared to a woman and advised her to eat from a tree. Even if we accept that snakes talk how do we account for what the reptile had to say. Consider the implications of this story. The talking snake appeared to know the mind of the lord. He knew that the Elohim lord was not allowing the man and woman to know their status in regard to the Elohim gods and was threatening them with death if they came into that knowledge. The serpent contested that eating the fruit of the tree of knowledge would not kill them but on the contrary it would awaken them to who they truly were. To know so much the serpent must have been a member of the Elohim team. If that were so all the Elohim may have been serpents.

The story of a talking serpent was one of the most telling in the Bible as it revealed the *zoology of God.* It seemed to suggest that the Elohim 'gods' were reptiles. Some readers may find the concept of God being a group of reptiles too shocking and far-fetched to contemplate. Nonetheless, it could account for the references to serpents and dragons in the Bible and in other scriptures, religions and mythologies.

The more primitive part of our brain is reptilian with a snake-like spinal column. A serpent is the nervous core of what we are. We could be viewed as serpents crowned by a cerebral cortex. Genesis could be telling us that we were bioengineered from reptiles. It may be saying we were predated by a humanoid reptile. Let's speculate around the fact that amphibians, reptiles and dinosaurs existed for millions of years before the advent of mammals and that life may not be restricted to the planet Earth. Prior to the appearance of the human being perhaps hominids evolved elsewhere in the Universe from reptiles. Maybe the Elohim emerged from dinosaur DNA. While the dinosaurs vanished on Earth elsewhere in the Universe they may have evolved into a humanoid type of creature.

The tempter of Eve may have been an advanced serpentine hominid. This might explain his ability to talk. If he was one of the Elohim then the Elohim story might have recorded the arrival on Earth of a team of reptilian extra-terrestrials. This would certainly help to explain why in so many parts of the world gods are portrayed as serpents.

This idea is in *Bringers of the Dawn.*[4] It is not a scientific or academic book but it contains information that appears to corroborate much of what I have been saying. Presented by Barbara Marciniak, the information it contains comes, supposedly, from humanoids on a planet orbiting a star in the constellation of Pleiades: *"Your history has been influenced by a number of light beings whom you term God. In the Bible, many of these beings have been combined to represent one being, when they were not one being at all, but a combination of very powerful extraterrestrial light-being energies. They were indeed awesome energies from our perspective, and it is easy to understand why they were glorified and worshipped. Who are these gods from ancient times...These beings were passed down through the ancient cultures of many societies as winged creatures and balls of light...The creator gods who have been ruling this planet have the ability to become physical, though mostly they exist on other dimensions...The gods raided this reality. Just like corporate raiders in your time...Before the raid, you had tremendous abilities...A biogenetic manipulation was done, and there was much destruction...your DNA was scattered and scrambled by the raiders a long time ago...Certain entities took the existing species, which was indeed a glorious species, and re-tooled it for their own uses...These creator gods set out to alter the DNA inside the human body...The creator gods are space beings who have their own home in space. They are also evolving...These space beings are part-human and part-reptilian..."*

If humans were originally bioengineered in the image and likeness of space beings, these *space beings* must have appeared similar to us thinking and acting much as we do, the only difference being that their bodies originated from reptilian stock rather than mammalian. In fact, we could be them with mammalian characteristics genetically spliced into their original reptilian DNA.

In our beginning we may have been a glorious creation of the Elohim subsequently retooled in a rebellion led by a rogue member of the group. That may be how the redaction to monotheism occurred. The talking serpent may have been a member of the breakaway group reneging on his rebel master; the one monotheists worship as God. There could be a lot more to the story of mankind than is revealed in the book of Genesis nonetheless our genetic connection to the Elohim is made clear in ongoing verses:

And it came to pass, when men began to multiply on the face of the earth, and daughters were born unto them, that the sons of God saw the daughters of men, that they were beautiful; and they took wives for themselves, of all whom they chose...and also afterwards, when the sons of God came into the daughters of

4 Marciniak B. Bringers of the Dawn, Bear & Co., 1993

men and they bore children to them (Genesis 6:1-4).

To come into women and have children by them the Elohim and their sons must have had species compatibility with humankind. This would have been possible if the Elohim bioengineered humanity by splicing genes from terrestrial mammals into their own species chromosomes. If, after upgrading their bodies from reptile to mammal, from the mammalian gene pool of the Earth, they reincarnated into the new human bodies they had formed, then most certainly they could have procreated with other humans.

If this sounds farfetched just consider the implications of extraterrestrial hominids being around for millions of years prior to humankind. If this were so it is possible they may have evolved on other planets to a point of high capability. Like us they may have developed advanced scientific and technical capabilities of genetic engineering.

The Earth is just a small planet on the outskirts of one amongst billions of galaxies. There could be trillions of planetary systems and amongst them many that are capable of supporting life. With billions of years to evolve in and millions of planets to evolve on, it is possible there are civilisations of intelligent life forms more advanced than us. From an abundance of circulating information, terrestrial science appears to be primitive by universal standards.

In my books, *The Vortex: Key to Future Science*[5] and *The New Science of the Spirit,*[6] *The New Physics of Consciousness*[7] and *AWAKEN* [8] and later in this book, I argue it may be possible to alter the speed of energy in the quantum vortex so matter can be changed from a state of physical energy to superenergy. This capability would enable bodies to move in and out of space time rather than through it. By this means advanced extraterrestrials could have overcome the perceived limitations of travel between the stars and visited our planet to bioengineer us.

The book of Genesis in the Bible makes sense if the Elohim were advanced beings who came to Earth from outer space. Genesis becomes plausible if they are viewed as the extraterrestrial bioengineers of humanity. This could also help provide an answer to the missing link in human evolution. Humanoid remains have been found on the Earth that are up to four and a half million years of age, yet leading geneticists state that the DNA for the planetary population of human beings can only be traced back to a common root about two

5 Ash D. & Hewitt P. The Vortex: Key to Future Science, Gateway Books, 1990,

6 Ash D., The New Science of the Spirit, The College of Psychic Studies, 1995

7 Ash D., The New Physics of Consciousness, Kima Global Publishing, 2007

8 Ash D., AWAKEN, Kima Global Publishing, 2018

hundred thousand years ago.

There are scant records because so much of the historical record was destroyed. Nothing survived the burning of the Egyptian library of Memphis. Two hundred thousand irreplaceable volumes were lost in the destruction of the library of Pergamos. The Romans incinerated five hundred thousand priceless manuscripts in the destruction of Carthage the and Christian Romans were responsible for destroying the greatest library in antiquity at Alexandria where an estimated seven hundred thousand hand written volumes went up in flames. The sacking of Constantinople caused the loss of yet more records from the past and what the Inquisition didn't destroy the Catholic Church has hidden from view in the Vatican vaults. Reliable records of extraterrestrial serpentine scientists tinkering with our genes may have been lost in these and other catastrophes.

Fortunately earlier records have surfaced that appear to confirm that visitors from space took command of the skies and interfered in our evolution. Archaeological discoveries in the 19th century uncovered cuneiform records on hundreds of thousands of clay tablets uncovered from the ancient Sumerian civilisation, which is where the book of Genesis appears to have come from. They provide overwhelming support for interpretation that the Elohim were extraterrestrials. Cuneiform Babylonian tablets in the British Museum describe the phases of Venus, the four moons of Jupiter and seven satellites of Saturn, none of which could have been seen from ancient Babylon.

The Turkish maps of Piri Reis, dated from the early sixteenth century are alleged to be based on earlier maps predating Alexander the Great. They accurately depict the Amazon Basin of South America and northern coastline of Antarctica, neither of which was surveyed until the advent of aircraft in the twentieth century. The coastline of Antarctica that has been under ice for several thousand years is clearly defined in these maps.

The Dogon tribe of Africa have an accurate knowledge of the Solar system. They claim their forebears received it from visitors from space.

Many people researching the ancient records support Marciniak's reference, to a raid by reptilian gods and the bioengineering of mankind. In his 1997 book, *The God Hypothesis*,[9] Dr Joe Lewels, former head of the Department of Journalism at the University of Texas at El Paso, gave the opinion that Jehovah was a reptilian being of flesh and blood who flew in craft and used a vehicle to transport Moses to the summit of Mount Sinai, as stated in the Bible: "*You have seen what I did to the Egyptians, How I bore you on eagle's wings and*

9 Lewels J., The God Hypothesis Granite Publishing 2005

brought you to myself. (Exodus 19:4)

Lewels also noted that Moses and the Israelites were never allowed to see Jehovah's face and wondered if his countenance was so nonhuman as to provoke fear and loathing. *"It should be pointed out that this is not in the least an original idea,"* wrote Lewels speaking of the Mandaens, an early Jewish sect who believed in a dualistic universe divided equally into the worlds of light and darkness. *"To them, the physical world, including the Earth, was created and ruled over by the Lord of Darkness, a reptilian being...variously called Snake, Dragon, Monster and Gian...thought to be creator of humanity."*

This same concept was advanced by researcher and author R.A. Boulay, who noted that practically all cultures of the world have stories of dragons or reptilians who coexisted with man – even created man – and were associated with powerful gems or crystals, walked on legs, flew in the air, fought over territory, and were revered by humans as gods. In *Flying Serpents and Dragons: The Story of Mankind's Reptilian Past,*[10] he wrote: "*The world-wide depiction of flying reptiles makes it abundantly clear that our creators and ancestors were not of mammalian origin but were an alien saurian breed.*"

In *The Gnostics* [11] Tobias Churton wrote: *There were Jewish schools, much given to speculation on the nature of God and the constituent beings which constituted his emanation or projection of being. Some of them appear to have been profoundly disappointed with the God of the Old Testament and wrote commentaries on the Jewish scriptures, asserting that the God described there was a lower being, who had tried to blind Man from seeing his true nature and destiny. We hear their echoes in some books of the Nag Hammadi Library, namely, 'The Apocryphon of John' and 'The Apocalypse of Adam. They believed in a figure, the 'Eternal Man' or 'Adam Kadmon' who was a glorious reflection of the true God and who had been duped into an involvement with the lower creation, with earthly matter, ruled by an inferior deity who, with his angels, made human bodies.*

Angels are extraterrestrial by definition.

The creation of Man, as described in the scripts of Genesis, has parallels in myths originating from cultures as diverse as those of the Tibetans, the Hawaiians, the Australian Aborigines, the North American Indians such as the Apache, Hopi and Sioux, the Maya, and on tablets found on the Easter Islands. These all speak of the involvement of space gods in the creation of man.

---

10 Boulay R.A. Flying Serpents and Dragons, Book Tree, 1997 (revised 1999)

11 Churton T. The Gnostics, Weidenfield & Nicolson 1987

The ancient Sumerian texts in cuneiform script on clay tablets are especially significant as they predate the Bible and the biblical patriarch Abraham was a native of Sumer. The knowledge he derived from that civilisation was passed down to his people and recorded in Genesis.

The Sumerian records, portrayed in the *Twelfth Planet* and *Earth Chronicles* [12] by Zecharia Sitchin, tell that about 450,000 years ago two half-brothers, Enki and Enlil, descending as gods from the heavens to the Earth. Sitchin emphasised the consistency of this claim: *"The statement that the first to establish settlements on Earth were astronauts from another planet was not lightly made by the Sumerians. In text after text, whenever the starting point was recalled, it was always this: 432,000 years before the Deluge, the DIN.GIR – [translated as] Righteous Ones of the Skyships – came down to Earth from their own planet."*

According to the Sumerian records, climatic changes, about 200,000 years ago, caused the DIN.GIR to create mankind as a hybrid of their own breed and stock of the Earth that would be better adapted to the new climatic conditions than their own workers. According to these legends we were a slave race. We were bred to mine gold. To begin with we were infertile mules produced in batches. Only later was our ability to reproduce switched on. Obviously from a modern perspective this could be interpreted as genetic engineering.

In *Rule by Secrecy* [13] Jim Marrs discusses the 500,000 clay tablets of cuneiform text recovered from excavations of the ancient ruins of the Sumer civilisation in Iraq and concludes the introduction of civilisation to the Fertile Crescent by extraterrestrials occurred some six thousand years ago. In his Eden series 1,2,3 Paul Wallis suggested that the book of Genesis was not detailing the creation of the Earth but an Elohim rescue and recovery programme after a global cataclysm, less than twelve thousand years ago, that left the Earth flooded and the skies darkened.

Dr Arthur David Horn, resigned as professor of biological anthropology at Colorado State University in 1990 after he concluded that the conventional explanations for man's origins he taught were nonsense. After much study he too came to the conclusion that extraterrestrials were intricately involved in the origin and development of humanity.

Swiss author Erich von Daniken, was harshly criticised by mainstream scientists and theologians, over his 1968 book on ancient astronauts, *Chariots of the Gods*,[14] but discoveries in archaeology and anthropology have reinforced his

12 Z. Sitchin, The Twelfth Planet, Bear & Co., 1991

13 Marrs J. Rule by Secrecy, HarperCollins 2000

14 Von Daniken, Chariots of the Gods 1968 (Souvenir Press new edition 1990)

theories. Harsh criticism was to be expected as *Chariots of the Gods* proved to be very popular. It sold over 40 million copies. In its day it was the top selling nonfiction book.

The extensive and detailed records of the Sumer civilisation that predate the Bible are not generally accepted because science is not impartial. Science is highly censored. In *Forbidden Archeology,*[15] Michael Cremo wrote: "*There exists in the scientific community a knowledge filter that screens out unwelcome evidence. This process of knowledge filtration has been going on for well over a century and continues right up to our day.*"

To quote again from *Rule by Secrecy* [13] by Jim Marrs: "*But woe to those who attempt to argue against conventional thinking...One particularly exasperated researcher wrote (Jonathan Starbright 1999) "Realize, that scientific institutions, such as the Smithsonian and the National Geographic Society, were set up by the world's elite factions in the first place to either debunk, distort or simply ignore any scientific data that tends to enlighten people of their true origins.*"

If the evidence of extraterrestrial involvement in the origin of our species is to be believed, then it is unlikely that life originated on Earth. Paul Wallis details how over time not just humans but other species were introduced to the earth by visitors from space. It is more likely that life is widely spread throughout the Universe. Unfortunately the theory of evolution does not support this view.

---

15 Cremo M. Forbidden Archeology Bhakti Vedanta Book Trust, 1998

CHAPTER 7
EVOLUTION

The scientific story for the origin of life is based on Charles Darwin's theory of evolution. His theory is an attempt by scientists to account for the origin of life in line with the philosophy of the Greek philosopher Democritus that everything in the Universe, including intelligence and life, emerged from the random interactions of material atoms. However, there are problems in this attempt to explain the origin of life, in particular the cell, in terms of chance interactions of inorganic matter.

The chance hypothesis for the evolution of life on Earth caught hold in 1953 when Stanley Miller passed electrical discharges through a mixture of gases replicating the early atmosphere of the earth. Amino acids were discovered in the consequent mix, which led the scientific community to conclude that Miller had discovered how life began. They seized on the idea that he had proved life emerged from lightning discharges which caused the formation of organic molecules from inorganic matter. In line with the predominant materialistic belief amongst scientists, the story that atoms charged by lightning jostled together to form amino acids which then formed proteins, morphed into scientific fact. Scientists argued that over billions of years these become organised by incremental steps into cells. This all happened by pure blind chance and the rest is evolutionary history, so to speak.

The discovery of amino acids in meteorites supported this story. If life was seeded by the arrival of organic molecules from space the possibility that life originated by chance was increased by the vastness of space. However, the Miller-meteorite-mythology is weak as scientists discovered that another molecule called RNA (ribose nucleic acid) is essential in the formation of proteins. RNA is made of small molecules called nucleotides, which are slightly more complex than amino acids.

In 2009 nucleotides were fabricated from electric discharges in a lab at Manchester University but this discovery did not explain how RNA first came to be formed. There is no evidence that RNA can form at random from nucleotides, just as there is no evidence that the appearance of amino acids in Miller's experiment could by pure blind chance lead to a fully formed protein.

Imagine we were wiped out in some terrible catastrophe and aliens landed on Earth outside the ruin of a cottage in the Peak District. Unable to explain how it came to be they began to search for clues. One of them found stones on the ground identical to those in the cottage walls and suggested the stones must have arranged themselves into the cottage by pure blind chance.

The alien said that over an enormous period of time the random organization of stones could have happened incrementally by the action of wind and wild animals. Another alien pointed out that the mortar holding all the stone together in the cottage walls was of the same substance as the stones. He wondered how it came to be set precisely between each stone and how they came to be perfectly positioned in the wall, with opening holes, just by chance. He reckoned it was beyond the bounds of probability that both mortar and stones could have become arranged together into the cottage walls by luck alone.

The stones represent amino acids in this story and the mortar depicts the nucleotides. The random appearance of amino acids and nucleotides through electrical discharges in gases is the equivalent of the aliens finding limestone near the cottage. Just as the random appearance of limestone near the cottage does not explain the cottage, so the random formation of amino acid and nucleotide molecules do not explain how these came to be arranged in the protein and RNA molecules observed in cells. The formation of organic molecules in electric discharge experiments did not prove they self-constructed by chance into complex molecules, or that they did it in incremental lucky steps over immense periods of time.

Protein molecules don't just self-construct. They consist of anywhere between fifty and several thousand amino acids that are organised by living cells in a very specific sequence that enables them to perform specialist tasks in the cells.

On average 200 amino acids are incorporated in a protein molecule. In every cell there are many thousands of different types of protein molecule. The chance of even one protein incorporating just 100 amino acids randomly forming on earth has been calculated at about one in a hundred thousand trillion. Multiply that by every protein that exists.

The sequence of amino acids in a protein molecule is set by the RNA molecule. RNA is required to form protein but proteins are essential to the formation of RNA so the proteins and RNA would have had to appear simultaneously perfectly formed for the first cell to come into being. The chance of that happening by pure blind chance is beyond the bounds of probability.

The alien who came up with the chance hypothesis went on to write a book called *The Blind Cottage Maker*. Though the book attracted a large following amongst the people on his home planet that didn't prove what he said in it was true. It just proved that most people don't think for themselves but go along with what cleverer people think rather than confront dilemmas in their own heads and challenge what they read and are taught.

The chicken and egg dilemma of life is that new life always comes from existing life. Every cell originates from a cell that has existed before it. Cells divide to reproduce but the big question is how did the first cell come into existence.

All living cells fall into two basic types; a more primitive type called *prokaryotic* and a more advanced type called *eukaryotic*. Eukaryotic cells have a nucleus. Prokaryotic cells do not. Most bacteria are prokaryotic. Most of the cells in our bodies are eukaryotic.

Many biologists believe that the eukaryotic cell is formed of a number of different types of prokaryotic cells, which came together to form its organelles. That is a sound theory for the origin of the eukaryotic cell, which fits well with the theory of evolution. The question remains, however, where did the first prokaryotic cell come from.

The most primitive prokaryotic cell is an extraordinarily complex thing. It is surrounded with a permeable membrane made of protein molecules which is many thousand times thinner than a sheet of paper but far more intricate.

Think of the cell as a medieval walled city. Just as the walled city had guarded gates so the cell membrane also has gates formed of protein molecules that select what can enter and what can leave the cell.

Some protein gates are tubes that allow the passage of only very small molecules such as air and water. Others are open at one end and closed at the other with the openings formed to fit a larger molecule of a specific shape. When a molecule with the right shape docks onto the opening the gate closes around it to allow its passage across the membrane, while the other end of the gate opens to allow it to enter or exit the cell.

The inside of the cell is full of watery liquid containing nutrients and waste as well as useful end products of the chemistry going on in the cell. Just as the walled city is full of workshops, the prokaryotic cell is full of sites where very specific chemical reactions occur. Nothing is haphazard. Every reaction is ordered and scheduled according to the needs of even the most primitive cell.

Most of the production is concerned with the formation of proteins from twenty or so amino acids. These are used to build the structures of the cell, such as membranes, and the enzymes which facilitate the complex chemical reactions. The enzymes are arranged with proteins and RNA to form a ribosome. The ribosome is the chemical factory where the chemistry of the cell occurs. These are equivalent to the different workshops in the walled city each designated to a different purpose i.e. the butcher, the baker or the candle stick maker.

Ribosomes make enzymes but they can only function with enzymes. How did the first ribosome form? The right amino acids required to form the right enzyme protein molecules must have come together, alongside an RNA molecule with the correct sequence of nucleotides. To imagine this occurred by pure blind chance is ludicrous because the math of probability does not allow for that degree of chance encounter. How could so many of the correct molecules come together and entirely by luck join together in the required perfect sequenced combination. It is a story that is hard to believe. We are left with another chicken and egg dilemma, which came first the ribosome or the enzyme.

In both types of cell, instructions in messenger RNA are received from another complex molecule called DNA (desoxyribose nucleic acid), which occurs in a central nucleus in the eukaryotic cell. The instructions in the DNA molecule determine the sequence of nucleotides in the RNA and the sequence of amino acids in the proteins.

DNA is a double helix which coils to form the chromosomes in the nucleus of a cell. Even though prokaryotic cells don't have nuclei they do have DNA double helix strands, delivering instructions from their binary code to RNA strands.

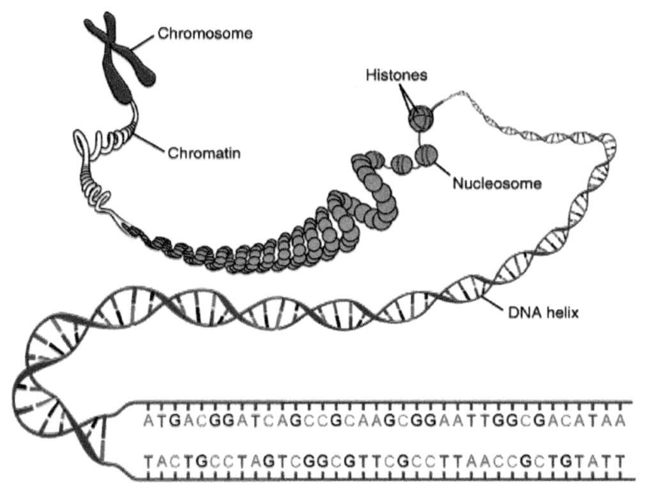

Yet another chicken and egg dilemma is raised by DNA. The formation and replication of DNA requires RNA and protein enzymes but the formation of RNA and enzymes requires DNA. The first cell to form would require DNA with the right sequence of nucleic acids to instruct the RNA before the emergence of the protein enzymes. If this happened by chance there is still the problem of sequence to overcome. The enzymes needed to form the DNA require the DNA to instruct the sequence of amino acids required in their formation. Which came first the DNA or the enzyme.

Finally there is the issue of energy. In the medieval walled city fire was essential for the baker to bake his bread, for the candle maker to melt his wax and for housewives to cook the butcher's meat. But the fire had to be controlled in the ovens to give precisely the right amount of heat.

In the cell energy is required to enable all the many and complex biochemical reactions to occur. Energy in the cell is provided in precise amounts by ATP (adenose triphosphate) operating in organelles called mitochondria. In the cell these organelles are the equivalent of the ovens in the medieval walled city. Enzymes are needed to produce ATP but ATP is required to produce the enzymes. Which came first the ATP or the enzymes.

Chicken and egg dilemmas abound in the cell. DNA cannot exist without enzymes but enzymes cannot exist without DNA. RNA needs proteins but to form proteins need RNA. Proteins are made in cells but cells are made of proteins. The simplest cell is made up of millions of proteins of thousands of different types. All these have to be working together within a selectively permeable membrane for DNA, RNA and ATP to operate.

Walled cities didn't appear on the landscape by chance. They were part of an evolutionary process involving intelligent people progressing from caves and mud huts to towns and cities. People with creative intelligence were behind the evolution of cities over centuries. Whereas some developments came about by chance none of them occurred by chance alone. The mud huts and cities were built by people. The existence of towns and cities is proof of the existence of intelligent people capable of designing and building them. Life screams intelligence but intelligence is not countenanced in the evolutionary theory for the origin of life because scientists have blind faith in the philosophy of materialism that underpins science.

To date scientists have catalogued some 200 million large fossils and billions of small fossils. Most researchers agree that this vast and detailed record shows that the major groups of all animals appeared suddenly and remained virtually unchanged. There is little evidence in the fossil record to support the theory of incremental evolution. In *The Triumph of Evolution and the Failure of Creationism* [1] Niles Eldredge admitted that over time little or no evolutionary change in species appears to have occurred. He said, *"Though I am a fervent believer in evolution and natural selection, I have to admit the evidence for natural selection is very weak."*

In *Sudden Origins* [2] Jeffrey Schwartz wrote: *Natural selection may be helping species adapt to changing demands of existence, but it is not creating anything new.* The theory that mutations drive evolution is not supported by the science.

In *Mutation, Breeding, Evolution and the Law of Recurrent Variation,*[3] Wolf-Ekkehard Lönnig, of the Max Planck Institute for Plant Breeding Research, wrote: *Mutations cannot transform an original species of plant or animal into an entirely new one. This conclusion agrees with all the experiences and results of mutation research of the 20th century taken together as well as with the laws of probability...properly defined species have real boundaries that cannot be abolished or transgressed by accidental mutations.*

So why do scientists promote the theory of evolution in schools, colleges and universities as scientific truth when it lacks supporting scientific evidence. Influential evolutionary biologist, Richard Lowentin, said this is because, "... *most scientists have a commitment to materialism and refuse to even consider the possibility of an intelligent designer.*" Nonetheless, there are a few scientists who are determined that evolution is driven by intelligence.

---

1 Eldredge N. The Triumph of Evolution and the Failure of Creationism, Freeman & Co., 2000

2 Schwartz J. Sudden Origins – Fossils, Genes and the Emergence of Species, John Wiley & Sons, 1999

3 Lönnig W. Mutation, Breeding, Evolution and the Law of Recurrent Variation, Recent Research. Development. Genetic. Breeding, 2(2005): 45-70 ISBN: 81-308-0007-1

CHAPTER 8
THE INTELLIGENT UNIVERSE

In *The Intelligent Universe*[1] the cosmologist, Sir Fred Hoyle, reasoned that evolution is driven by intelligence. Hoyle considered it a vast unlikelihood that life could have evolved on our planet alone and from non-living matter without involvement of intelligence. In his words: "*...it is apparent that the origin of life is overwhelmingly a matter of arrangement, of ordering quite common atoms into very special structures and sequences. Whereas we learn in physics that non-living processes tend to destroy order intelligent control is particularly effective at producing order out of chaos. You might even say that intelligence shows itself most effectively in arranging things, exactly what the origin of life requires.*"

A process of evolution driven by intelligence is evident in our daily lives. We put out ideas and see if they work. If they don't we adapt them until they do or drop them and try something different. Evolving theories are subjected to criticism so that only the best survive. But all the time we are in the process. The work would never happen without our intent and our conscious involvement learning from a process of trial and error. Consciousness, intent and intelligence are never excluded from the human creative process. As we take rejection in our stride and learn from our mistakes, we modify and make choices. There are lucky breaks but for the most part it is a learning process,

---

1 Hoyle F. The Intelligent Universe, Michael Joseph, 1983

taken step by meticulous step. In *The Blind Watchmaker*,[2] Richard Dawkins suggested that the process of evolution occurred as a series of lucky incremental steps but his description was not based on experimental science, neither did it preclude intelligence.

If, as Hoyle suggests, the Universe is intelligent then life on Earth could be the culmination of a process of intelligent evolution on a universal scale. Maybe we are asking the wrong questions. Debating whether life is a product of mindless evolution or designer creation would be a pointless exercise if life was shown to be an expression of intelligence innate in the Universe. If the Universe is intelligent, it would be illogical to exclude intelligence from evolution.

Life screams intelligence. Research is revealing the innate intelligence of plants and animals. Slime mold appears to be as smart as Japanese engineers and apes can solve mathematical tests faster than humans. That suggests it might be a bit arrogant of us to hog intelligence and exclude it from evolution. There is no doubt Darwin's theory of evolution is brilliant and neo-Darwinian thinkers have made contributions but perhaps it might be time to allow for intelligence in the theory of evolution.

In the quantum vortex theory I use evidence from high energy physics to support the premise that there is no material substance underlying matter and that particles of energy are forms of non-substantial motion. As such they appear to be abstractions and the quantum appears to be more in the nature of a thought than a thing.

Mind is a body of thought. If the quantum of energy is treated as a thought then, as a body of energy, the Universe could be described as a mind. As mind and thought are synonymous with intelligence the quantum of energy could be described as a particle of intelligence. If that were so, intelligence would be in the quantum fabric of the Universe.

Intelligence is synonymous with life. If particles of energy are more like thoughts than things intelligent life could exist at every level in the Universe from the quantum to the galaxy. There would be no such thing as inanimate matter if particles of intelligence underlie everything.

If the Universe is constituted of quantum intelligence there would be no need to assume an intelligence outside the Universe responsible for its creation. If the Universe is innately intelligent it would be equally senseless to exclude intelligence from the evolution of life.

If the Universe is intelligent, evolution would make sense as a means whereby

2 Dawkins R. The Blind Watchmaker, Norton & Co, 1986

universal creative intelligence could have developed the multitude of forms that populate life. An intelligent Universe would never stop evolving, learning and growing, breaking things down and then building them up again better than before. The extraordinary diversity and dazzling beauty we witness in life doesn't require an external creator that made everything perfect to begin with. The evolution of life on earth could be an expression of the intrinsic nature of the Universe endlessly self-creating, All this is possible if intelligence is innate in the quantum underpin of the Universe.

Charles Darwin may have discovered aspects of the process by which universal creative intelligence develops life. He didn't discover there is no creative intelligence underlying life. One of Darwin's most brilliant insights was natural selection. Natural selection reveals mistakes. It is never easy for an author to spot errors. Authors rely on editors and critics to point out the flaws in our work. Life is harsher. Anyone not up to par ends up as lunch.

Chance is vital in evolution. Random events are essential to creative evolution because they enable freedom, and freedom of thought is essential to creativity. I think Einstein was wrong when he said *"God doesn't play dice."* I believe it is through chance that Life plays. The Western mind fails to see the underlying order in apparent chaos and the perfect operation of chance. Not so the Chinese mind. In the words of Carl Jung, from a foreword to the *I Ching*: [3] *The Chinese mind, as I see it at work in the I Ching, seems to be exclusively preoccupied with the chance aspect of events. What we call coincidence seems to be the chief concern of this peculiar mind and what we worship as causality passes almost unnoticed.*

In intelligent systems the freedom of random events and unforeseen problems set up endless opportunities for creative originality. Evolution may be witness to universal intelligence operating in biological systems, building things by trial and error watching them perform then breaking them down in order to reconstruct them in a better way. This is the eternal cycle of creation, preservation and destruction – birth, life, death and then rebirth - depicted in the mysticism of India as the trinity of creator, preserver and destroyer. People in the East do not preclude chance from their creation myths, they see chance as footsteps of the divine.

In the chaos of random molecular interactions chance may have a vital part to play in the evolution of life but it is not necessarily the driver of evolution. Fervent belief in pure blind chance, as the sole driver of evolution, is a lingering legacy of classical materialism that is in denial of the divine.

---

3 Wilhelm R. I Ching, Routledge & Kegan Paul, 1951

The blind chance hypothesis in evolutionary theory originated in the ancient Greek civilization. It came from the atomic hypothesis attributed to Democritus but the *a-tomos* meaning that which cannot be cut was not his original idea. Democritus was taught that the Universe is essentially granular and everything in it is derived from chance encounters of indestructible material atoms by his teacher Leucippus. Whether the philosophy of materialism began with Leucippus or was taught by his master we will never know. All we do know is the atomic hypothesis is still the bedrock of science despite evidence to the contrary in quantum physics

Most scientists would rather turn their backs on experimental science than ditch materialism. In this regard these scientists are no better than religionists in their adherence to blind faith. This is especially true in regard to the ongoing debate about creation versus evolution.

The battle between evolutionists and creationists is fundamentally a battle of faith. Neither camp is secure in experimental evidence. Both rely on conjecture. We could walk away from this battle of beliefs and look at the scientific evidence in an unbiased way.

If we accept the evidence from quantum physics that everything is energy and energy has no substance, logic leads us to premise that particles of energy may be more in the nature of thoughts than substantial things. That, in turn, leads us to presume that the Universe is more a mind than a material reality. If we accept that we would be led to join many quantum physicists in concluding that consciousness is the bedrock of reality.

CHAPTER 9
CONSCIOUSNESS

You have just woken from sleep and look into the mirror first thought you have is, "Oh what a mess I am in this morning". The great debate is who is thinking that thought. Is it the body thinking about itself or is it an individual psyche in the body thinking about how the body vehicle looks.

A devotee of Democritus would say that the body thinks about itself through atomic interactions in the brain. Their contention would be that mind is just a manifestation of matter, a consequence of biochemical reactions in brain cells.

Someone belonging to the school of thought, attributed to Plato, would say that a psyche is thinking through the body about the body. Their contention would be that mind exists separate from matter but by incarnating in a body it is able to express itself in the realm of physical matter through the biochemistry of the brain.

After breakfast you go out to the car. You look in the rear-view mirror and think "Oh what a mess my car is in this morning." Would you say the thought about the car was coming from someone in the car thinking about the car or would you say the thought was coming from the car itself? When you are in the car you are an integral part of the car. It cannot drive without you and you cannot cover long distances quickly without it. However, when you are in your car thinking you do not imagine it is your car thinking. If you killed someone by dangerous driving the judge would not send your car to prison no matter how vehemently you might argue your car was responsible.

Consciousness is said to be the hard question. Does consciousness originate in the brain or is the brain derived from consciousness. Is consciousness the consequence or the cause of physical reality. Albert Einstein believed that space and time go back to a start point of pure consciousness. Some two thousand four hundred years before him Plato taught the Universe is the outpouring of a single source consciousness that brings everything into being by pure intent.

Plato held the position that consciousness animates the body, coming into it at birth and leaving it at death. But when Plato was a boy, Democritus was teaching that consciousness is not an independent reality. He said it arises from the animation of atoms in the body. Consciousness is only a hard question for people who believe that consciousness arises from molecular activity in brains and expires when brains die despite NDE evidence that reveals consciousness survives death of the brain.

If the properties of material particles are explained away by vortex motion and the material philosophy of Democritus is revealed as spin, scientists and people of a scientific bent should be able to grasp the quantum concept that particles of energy are not substantial things and appear to be more like thoughts. If they can accept that then because thoughts require consciousness, they may be able to go beyond the idea that consciousness is the consequence of brains and think of it as something more universal. Then they might conclude that consciousness is not derived from matter but matter is derived from consciousness. That would bring them in line with Max Planck, the father of quantum theory, who is purported to have said, *"I regard consciousness as fundamental. I regard matter as a derivative of consciousness."*

If people want to know where God fits into the equation, quantum physicist, Amit Goswami, suggested that God is the universal consciousness at the ground of all being. In *The Self-Aware Universe,*[1] he wrote *"You can call it God if you want, but you don't have to. Quantum consciousness will do."*

In an interview for the Observer, following publication of *'The Mysterious Universe',*[2] Sir James Jeans said: *"I incline to the idealistic theory that consciousness is fundamental, and that the material universe is derivative from consciousness, not consciousness from the material universe... In general, the universe seems to me to be nearer to a great thought than to a great machine."*

There cannot be a great thought without a great thinker and the argument for universal consciousness being the great thinker begins with quantum physics. In physics energy is defined as a measure of activity. In material physics energy was treated as a measure of the activity of material particles. But quantum physics has established that all particles are forms of energy and energy has no material foundation; it has no substance. The quantum vortex has explained away the properties of material substance as spin. Energy cannot be an act of material substance if material doesn't exist. If energy is not the act of material substance what else can it be but act of consciousness. Many quantum

1 Goswami A. The Self-Aware Universe: How Consciousness Creates the Material World, Tarcher Pedigree, 1993

2 Jeans J. The Mysterious Universe, Cambridge University Press, 1930

physicists believe that consciousness underpins energy, that it is the ground of all being and universal consciousness is the great thinker.

The Universe, is a collective of many particles of energy, but it appears to be underpinned by one consciousness. The singularity of consciousness is clear in physics. Every proton has precisely the same mass and the same value of electric charge, magnetism and gravity as every other proton. If protons are vortex particles of energy existing as acts of universal consciousness their identical properties suggest that they originate from a single consciousness. While the protons are many the consciousness underlying them all would seem to be one. As protons constitute most of the mass of the Universe this logic leads to the conclusion that there is a single source consciousness underlying everything, including you and I. If this is true the implications are immense. Those who believe in God tend to imagine God as an almighty Creator of the Universe in which we are just an insignificant part. But if God is universal consciousness it could not be presumed there is a big God consciousness up there imagining the Universe into existence while we are miniscule bits of human consciousness down here wondering why we exist. That which is singular cannot be divided so if we are conscious the consciousness within us has to be the same one consciousness that brings the Universe into being.

The indivisibility of consciousness suggests that while we are separate in our bodies we are not divided in our consciousness. The implication of this is that humanity is a single universal consciousness in many different bodies. Our life experiences might be many but that within us which is aware of our individual life experience appears to be one. It could be we are one being in many bodies.

The evidence of NDEs suggests that you are not your body. NDEs indicate you are conscious awareness looking out of the eyes in your body, hearing through your ears, smelling through your nose, tasting through your tongue, feeling through your skin. According to the evidence of NDEs when you die your conscious awareness separates from your physical body and this real you goes on, fully conscious and aware, into another dimension of reality. What I am suggesting goes a step further. I am suggesting that the essential you, the consciousness that breaks free of physical incarceration is the same one consciousness that is holding the new world you enter in being. If consciousness is one and indivisible and it is the same one consciousness in you and in me and in everyone and everything else, then at this very moment you and I and everyone else may be the same one being that is maintaining through conscious intent, the Universe and every dimension in it. NDE's suggest you are not your brain. It is not the source of your consciousness. It puts a limit on your consciousness. This view was put forward by Aldous Huxley in *Doors of Perception* [3] People who have a transcendent experience and are aware of themselves as universal consciousness often share this view.

Plato taught that the Universe came into being so that consciousness can express and experience itself. He also taught that each one of us is a unique expression of the one consciousness. The theory of the quantum vortex supports his view that we are the means whereby the Universe can experience and express itself. If we are the same one conscious being in many bodies then if we hate and hurt others we hate and hurt ourselves but if we love and care for others we love and care for ourselves. The brotherhood and sisterhood of humankind are embodied in the theory of the single universal consciousness underlying everyone.

Paul Wallis concluded *Echoes of Eden* [4] with the statement: Look deep within yourself and you will find the Universe. Peer deep into the Universe and you will find yourself. Buddha encapsulated this principle in his Lotus Sutra: "Our existence is identical to the Universe as a whole, and the Universe as a whole is identical to our existence." So important is this mystic law of life – We are in the Universe and the Universe is in us – that Nichiren Daishonin, in thirteenth century Japan, encouraged people everywhere to chant, ***Nam-myoho-renge-kyo,*** which translates as: Remember the mystic law of life revealed in the Lotus Sutra. Today millions of people throughout the world - including myself - chant this affirmation of the oneness of humanity, which helps to bring unity and harmony to humanity as a whole and peace, prosperity and wellbeing to individuals, families and relationships.

If a single, indivisible consciousness underlies everything this unity of being would also apply to plants and animals and even inanimate things like rocks, the soil and the sea. The implication of this is that if we are cruel to the planet and to plants and animals we are being cruel to ourselves but if we care for plants, animals and the planet we are caring for ourselves. Our survival and the survival of the planet depends on our recognising this mystic law of life, which is a core spiritual belief.

---

3 Huxley A., The Doors of Perception, Flamingo 1994

4 Wallis P., Echoes of Eden, Paul Wallis Books, 2022

CHAPTER 10
GOD AND ENERGY

In ancient India yogis recognised there was an ultimate creative principle underlying everything. They called it Brahma. In science the ultimate creative principle underlying everything is called energy. In Christianity the ultimate creative principle underlying everything is called God.

Brahma and God are attributed with consciousness and intelligence but the difference between Brahma and God is that in Christian theology God is placed outside of the Universe whereas in the Indian tradition of Yoga, which means union, there is no separation. In Indian philosophy Brahma is not separate from the Universe. Brahma is at one with the dynamic state we call energy. There is a hint of commonality between the dynamic state we appreciate as energy and the source commonly called God. It is apparent in the definitions of energy in science and God in religion.

- Energy is neither created nor destroyed
- God is neither created nor destroyed
- Energy is everywhere
- God is everywhere
- Energy is in everything
- God is in everything
- Energy is all might
- God is almighty
- Everything comes from energy
- Everything comes from God

In the Bible, the Gospel of John opens with the verse: *"In the beginning was the Word and the Word was with God and the Word was God."* Word is sound. Sound is vibration. Vibration is energy. So the first verse of John's Gospel could read: *"In the beginning was the Energy and the Energy was with God and the Energy was God."* This translation of the first verse of the Gospel of John suggests unity between God and energy. If God is defined as universal consciousness and energy is the expression of consciousness this verse supports that union.

However, while God and energy may appear to be in union there seems to be a distinction implied by the Word being God but also being *with* God. This distinction is also in the idea that particles of energy are more thoughts than things. If the universal consciousness underlying everything is the thinker and particles of energy are the thoughts then it could be argued that the thoughts are not the thinker, they are expressions of the thinker. Believers in God might say energy is *of* God it is not God. So where does the concept of simultaneous union combined with distinction come from? It comes from the non-substantial nature of reality.

Thoughts are not things. Thoughts have no material substance. If we believe there is a Great Thinker imagining the thoughts that constitute the universal mind we cannot assume that the Great Thinker is substantial. This is because substantiality has been dispelled as an illusion. Without substance the Great Thinker would exist only by virtue of thinking. If it were not a substantial being thinking what could it be but the state of thinking or the capacity to think. If the capacity to think ceased to think it would cease to be, as would the thought. Thoughts depend on the thinker for their existence but a non-substantial thinker would also depend on a continual stream of thoughts for its existence. Thinker and thought would always coexist. They would be in a state of total codependency. They could not be separate. In yoga or union they would be two aspects of the same codependent reality. The thinker would be the internalised source of thought and the thought would be the externalised expression of the thinker. When we drop the idea of substantial being we are left with the idea of states of being.

If we break away from the anthropic projection of a substantial being with a capacity to think we are left simply with the capacity to think being the state of being. The traditional idea of God as a sentient being with a capacity to think is a projection of our perception of ourselves as sentient beings with capacities to think. If the one universal mind exists purely as the capacity to think it would be simultaneously the thinker *of* all sentient beings and the capacity to think in all sentient beings. We are conscious and we have the capacity to think and be aware. If God is the capacity to think and be aware then our capacity to think and our conscious awareness would be God.

To think a Universe into existence would require more than conscious awareness. It would require intent to imagine every quantum vortex and quantum wave-train of energy into being. We cannot begin to comprehend that level of thinking but we see it operating in ourselves. We all have the willpower of intent, also called, *free will*. We can intend things into being. This is a capacity attributed to God. We can create and creation is attributed to God. We can love and God is love. These are attributes of the *All That Is,* which include consciousness and thinking, joy and life, reason and ingenuity. All these values, expressed externally by finite human beings, are attributes of the infinite mystery called God residing in each one of us.

Once we abandon materialism, instead of imagining God as a being lording over us as miserable little beings, we could imagine God as the inner divine attributes of loving, living, joy and feeling we share in the commonality of being human with every other human being. From that perspective the one God would be what we are internally and the many bodies of mankind would be what we are externally.

In the Bible it is written: *Be still and know that I AM God* (Psalm 46:10). If God is our I AM presence, the essence of who we really are, then we would be God experiencing and expressing as the one life in many human beings.

If God is treated as life everything begins to makes sense. One life is common to us all and to all sentient beings. No one can say for sure what life is. From NDEs we know human life survives death. It is a breakthrough in our understanding of life to appreciate it survives death. NDEs help us appreciate life as a ubiquitous principle that animates all organisms and brings them into existence, sustains them and then destroys them only to regenerate them all over again. Hindus appreciate creation, preservation and destruction leading back to creation as the endless cycle of life attributed to God.

If the capacities we call God come from within us, it would make sense to seek God not outside of ourselves but within ourselves. This is the purpose of meditation. It is the way of yoga. It is the intent of the Buddhist chant, ***Nam-myoho-renge-kyo.***

As human beings we epitomise the relationship between God and energy. What is internal to us, our conscious awareness and our ability to think, create, love and reason, and our intent or free will, is our inner God aspect. What is external to us, our individual bodies, our genetic predispositions and our soul personalities is our energy signature. Taken together what is external and what is internal makes us the sentient human beings that we are. Hu is an ancient Egyptian word for God or the *Divine.* Hermes, was an Egyptian teacher who founded the ancient Greek civilisation three generations after Moses. He taught that we are divine beings in mortal bodies. This is why he called us *human beings.*

In the Copenhagen interpretation of quantum reality, reality is taken to exist only to the extent it is observed. Quantum particles of energy exist only to the extent they are observed by conscious awareness. If materialism is abandoned – the philosophy that God or the atom pre-exist activity – then conscious awareness cannot be assumed to pre-exist the quantum it is aware of.

In philosophy pre-existence is a problem. 'Imagine a creator in an eternal void before the Universe began thinking, *"Hmm today I will create a Universe!"* There is a 'pre-existence problem' with that idea. Space and time belong in the Universe. Properties of the Universe cannot be attributed to anything existing before the Universe began. Voids and beginnings are attributes of space and time. They can only occur within a Universe that already exists. The proposition that they occurred before the Universe existed, before space and time began, is a logical absurdity. The idea that God existed before the Universe makes no sense in relativity physics. If time is relative to the speed of light it cannot pre-exist energy. If time is a measure of the relationships between entities in the Universe it cannot exist without those entities. The Universe cannot have a beginning or an end. It could not have begun some thirteen and a half billion years ago. If there was a big bang it must have been preceded by a big crunch.

Think about it, the protons that make up the Universe are trillions of years older than the Universe itself. Protons have been around for at least a billion, trillion, trillion, years and that is just an estimate. As spontaneous proton decays have never been observed - though scientists are constantly looking for them - protons could have been around for a lot longer. Until a spontaneous proton decay is observed they can be presumed to be eternal. How can the particles that make up the Universe be trillions of years older than the Universe itself. There are serious metaphysical problems in regard to believing in a God that preceded the Universe, or believing the Universe began with the big bang. It is more likely the big bang was just one of a many phases of universal expansion and contraction. Sages in India speak of a periodic expansion and contraction of the Universe. They call it the breath of Brahma.

Yoga is union. Yoga is the union of the inner realty of consciousness and the outer reality of the observed Universe of energy. In yoga there is no separation between energy and consciousness. As my daughter Jessica - a teacher of yoga teachers - said, *"Everything goes back to Yoga."* If we tread the path of yoga we will come to know, from personal experience, that there is no division between God and the Universe. We will appreciate from our inner knowing the union of God and energy. This is the Gnosis. Yoga in the East and the Gnosis in the West are both concerned with our coming into the knowledge of our inner divine self which is the self of everyone else and the source of the Universe.

CHAPTER 11
FIELDS OF CONSCIOUSNESS

There is another form of matter apart from the atom which could impact our understanding of life and its origins. Only 0.01% of the Universe consists of atoms. 99.9% of matter is non-atomic. Most quantum vortex particles of matter are not in atoms, they occur in a form of matter called plasma. Around 98% of the mass of our solar system is plasma in the sun. Combined, the planets orbiting the sun that are made of atomic matter add up to 2% of the mass of the solar system.

In an atom electron vortices orbit a central nucleus of proton and neutron vortices of energy and the vortex energy, extending between the electrons and protons, sets up a force of electric charge attraction that holds the electrons in orbit.

In most atoms, the oppositely charged particles are paired off so that the atom is electrically neutral. If life depends on electricity, atomic matter would be the worst stuff in the Universe for supporting life. However if electrons are added to an atom or taken away from it then the atom becomes an electrically active ion. In atomic matter ions can support life because they can conduct electricity.

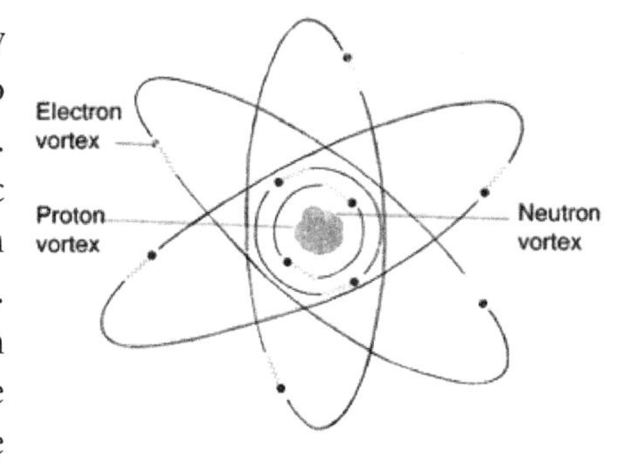

In plasma, electrons don't orbit atomic nuclei. As electrons and protons are not paired in plasma it is electrically active and therefore it is ideal for generating conditions favourable for life.

The Universe is mostly plasma, which is capable of conducting electricity, so if electricity is essential for life the Universe could be alive; it could be a living organism. It could be a reasoning sentient being like you and I.

Most plasma in the Universe exists in stars. Stars are nuclear furnaces where temperatures are too high for atoms to form. Electrons are racing too fast in stars to orbit protons and atomic nuclei. Instead electrons, protons, neutrons and atomic nuclei move about at high speeds in total chaos. But stars continually lose plasma into space. The sun is a star and as well as radiating heat and light it continually emits high energy electrons, protons, neutrons and atomic nuclei. These *cosmic ray particles* pour out of the sun into space as *solar wind.* The sun also loses plasma into space as coronal mass ejections - known as *solar flares.* This is true for all stars. All stars blow off plasma into space. The consequence of this is that space accumulates plasma like the seas accumulate salt, and there is a direct corollary here. The salty water in our bodies is called plasma.

In stars plasma is hot but in space plasma is cold. The cold plasma in space may not be chaotic so it might lend itself to organisation into electric field formations, which could set up conditions favourable for life. Let me explain.

If the nucleus of an atom were the size of a golf ball the nearest orbiting electron would be about two miles away. Moving freely in space, electrons, protons, nuclei and atomic nuclei would be much further apart. Though the distances between charged vortex particles in space may be vast, because their electric and magnetic fields extend into infinity they could form electric and magnetic fields and these fields could be organised and transmit information throughout the Universe.

If space contains charged quantum vortex particles it could be full of fields of electric activity. If electric activity is the basis of life then these fields could be alive to the extent that they could conduct electricity and transmit information. These fields of plasma could transmit conscious information and provide a physical base for conscious awareness. If that were so electric and magnetic fields of plasma in space could set up electric *fields of consciousness* throughout the Universe. These ideas are speculative but they could make sense of spiritual beliefs and give us a better understanding of spirits, angels and gods.

The basis of spiritual belief is that there are invisible, low density beings in space called *spirits.* It could be that spiritual life is a perception people had of the fields of consciousness. If plasmic fields of consciousness exist in space they would be invisible because cold plasma does not reflect light. Hot plasma in the stars and sun emits light, which is reflected by atomic matter. This enables us to see it. Non-atomic plasma is only visible if it emits light. Some forms of cold plasma emit light and are luminous. Most forms of cold plasma remain invisible as they do not emit light.

The definition of spirits is that they occur in space, they are conscious and alive and they are low density and invisible. Fields of consciousness fit this definition precisely.

There are many forms the plasmic fields of consciousness could take. Detected through its electromagnetic properties plasma has been observed to form vortices or helixes and vast filament networks connecting the stars. Cold plasma occurs in spheres called orbs. These have been detected by electric apparatus. Orbs can be suspended. Sometimes they move about. Invisible to the naked eye orbs are often caught as images on electrically operated digital cameras. People say they are prevalent in places where the presence of spirits is felt. These orbs could be cold plasma fields of consciousness formations. They could be spirits.

The fields of consciousness could form non-biological, non-atomic life forms in outer space and in spaces in our world. Space could be full of non-biological life. Some scientists believe this to be so. The author, Mark Heley, reported the observations of a Russian scientist that could impact our understanding of life.[1] *V.N. Tsytovich, a scientist at the Russian Academy of Sciences, has shown how plasma can self-organise when exposed to electric charge. Tsytovich has developed his observations into a theory of inorganic life. The principal location of this is in the helical dust structures that have been seen to form around stars and in interstellar space.*

In a gravity free environment, these plasma particles bead together to form string like filaments, which then twist into helix-shaped strands closely resembling DNA. These structures are electrically charged and are attracted to each other. They are able to 'feed' by assimilating other less organised plasmas through their boundary walls. They can 'reproduce' by amoeba-like splitting, and each of the plasma's offspring retains the capacity for self-organising growth, and further re-production. According to Tsytovich "they are autonomous, they reproduce, and they evolve," behaviour fulfilling enough criteria, in his opinion, to be considered a form of life.

If there are forms of plasma that are alive, these would be lighter, less solid and more fluid than the atomic bodies we inhabit because in atomic bodies electric fields lock atoms together to form solids, liquids and gases. None of these atomic constraints would apply to plasmic life forms; the simplest of which would be a sphere.

Living forms of plasma could sense each other and their spatial environment

1 Heley M., The Everything Guide, Adams Media. 2009

by feeling the forces of electricity and magnetism that lie between them. This is how they could be conscious and aware.

If spiritual matter is plasma and physical matter is atomic then the difference between physical and spiritual life could be the difference between atomic matter and plasmic matter. The plasmic theory for spiritual reality could pave the way for research in physics into spiritual phenomena. We could begin scientific lines of enquiry into the links between natural and supernatural life by thinking of life in terms of electrical activity and spirits as electric and magnetic fields of consciousness.

In religion both God and angels are defined as *spirits* so there could be a confusion between God and angels. This confusion is clear in the Bible in a communication between God and Abraham when he was called to sacrifice his son. The angel that appeared to stay Abraham's hand spoke as God: *Then the angel of God called again to Abraham from heaven, "I, the Lord..."* (Genesis 22:15) In this verse God reveals himself as an angel.

So far I have only spoken of plasma in terms of energy. But just as atomic matter can exist in the form of physical energy and superenergy, the same could be said of plasma. As well as physical energy plasma, the Universe may be full of superenergy plasma.

If the Universe were connected by electric fields of consciousness, treated as living entities growing and evolving, and transmitting and receiving information, the collective of these components of intelligence could be thought of as God. The contributions of this collective of intelligence to the evolution of life on Earth could be construed as creation by God. On a universal scale God could be thought of as a universal brain or computer functioning as the whole made up of we and the plasmic spirits as the parts. If we were *fractals or holograms* of this universal pattern it could be said that we are made in the image and likeness of God.

This description of us as a group consciousness, involved in the creation of the Universe, is a very empowering idea. If we are incarnations of the plasmic fields of consciousness that are responsible for creation then in our essential being collectively we could be the living conscious, electric fields of conscious intent underlying everything in our world, animate and inanimate. If we believe that story we would have reason to drop victim mentality and take personal responsibility for everything that happens to us and everyone else.

CHAPTER 12
WATER AND LIFE

We know that life on earth depends on water, but why? What is it about water that enables it to support life? The answer may lie in its ability to conduct electricity. Electrical conductivity depends on charged particles. Most atoms are electrically neutral but they can become electrically active when they gain or lose electrons and become ions. Metals, graphite and water can also conduct electricity.

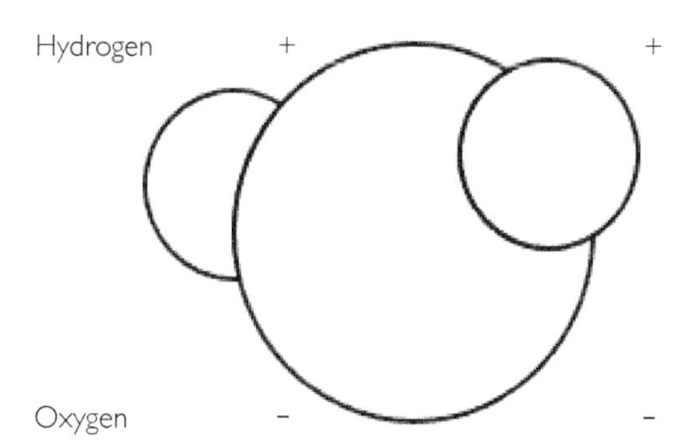

Water is an oxygen atom combined with two hydrogen atoms. In this combination the electrons belonging to hydrogen are predominantly in the orbit of the oxygen because oxygen atoms are electron hungry. The result is electrons are more likely to be found around the oxygen atom of the water molecule than the two hydrogen atoms. The two hydrogen atoms act as 'horns' of positive charge while the oxygen acts as a 'belly' of negative charge. This electric dipole of water sets up an electric force of attraction between one water molecule and another. A weak 'belly to horn' bond occurs between them, which is called hydrogen bonding.

Hydrogen bonding and the dissociation of some water molecules into charged ions enables water to conduct electricity and dissolve salts. As they dissolve the salts release charged ions into the water. That greatly increases its electric conductivity. This could explain be why salt water is the medium in which living organisms first formed on Earth.

Biological life depends on the electrical conductivity of salty water. In salt water ions are called electrolytes. In living organisms salty water is called plasma. The fluidity and electrical activity of watery plasma may be the nearest thing on Earth to electrically active cold plasma in space. If life depends on electric activity it would make sense that a planet containing an abundance of salt water would be ideal for establishing life in atomic matter. This may be why biological life emerged in the salty sea.

The biological cell is essentially a drop of salty water suspended in a gel of protein. The electrical activity of life in the cell, made possible by the salt water, was detected as an electric field around cells and multicellular organisms by an emeritus professor of medicine at Yale University.

In the 1930s, Professor Harold Saxton Burr, measured weak electric fields in and around living organisms. He named these the *Electrodynamic Life Fields or L-fields*. His research was extensive. Burr published around thirty papers in which he presented evidence that the subtle L-field has a major impact on the development and morphology of living organisms.

In his book on the life field, *Blueprint for Immortality*[1] Burr wrote: *When a cook looks at a jelly mold she knows the shape of the jelly she will turn out of it. In much the same way, inspection with instruments of an L-field in its initial stage can reveal the future 'shape' or arrangement of the materials it will mold. When the L-field in a frog's egg, for instance, is examined electrically it is possible to show the future location of the frog's nervous system because the frog's L-field is a matrix which will determine the form which will develop from the egg.*

The electrodynamic L-field has holographic properties in that every portion of the Life-field contains the blueprint of the entire organism. This is displayed in the early stages of its development. If the cells of an embryo are divided in half each half will develop into a completely formed organism. This occurs with

1 Burr H. Blueprint for Immortality, Neville Spearman, 1972

identical twins. Just as the division of a hologram results in complete images so the division of the embryo results in identical twins. Like the hologram every part of the L-field - also called the biofield - is a blueprint for the whole.

In her book, *The Field,*[2] Lynne McTaggart cited a number of scientists, associated with universities in the USA and France, Germany and Russia, who discovered a close association of electromagnetic fields with life. The question that goes with these discoveries is how these weak electric fields could overlay and inter-penetrate a body of atomic matter when very strong electric fields occur inside atoms between orbiting electrons and protons in the nucleus. How can the electric fields of life operate over these powerful electric fields. The subtle electric fields of life could only influence atomic matter if they managed to avoid direct interaction with the electric fields within atoms and between atoms in molecules. That might be possible if the L-fields were formed of superenergy.

Superenergy can interact with forms of matter in the physical world of energy without interference from physical energy fields. This is because physical energy does not interact with superenergy as superenergy exists outside of the space and time of physical energy, while superenergy can interact with physical energy because physical energy exists within the space-time continuum of superenergy. The principles of energy subsets and coincidence would make it possible for a field of consciousness in superenergy to overlay a biological cell. With no interference from the physical electromagnetic fields, superenergy electromagnetic fields could transfer vibrations to a body of physical energy. These vibrations could transmit information into the physical dimension through resonance.

Vibrations can be transferred between matching forms by resonance. For example, resonance occurs between a sounding tuning fork and a piano wire tuned to the same frequency. In the interface of coincident matching forms in the dimensions of superenergy and physical energy there is a possibility resonance may occur. Resonance occurs between tuned coils in a radio sets and broadcast electromagnetic waves. This is also how televisions work. Resonance between vibrating waves of superenergy and coils of physical energy could enable the transfer of information between the dimensions through DNA because DNA is coiled in the nuclei of biological cells. I believe the coil structure of DNA could enable DNA *resonance* to occur between the fields of consciousness and living organisms to drive evolution.

---

2 McTaggart L. The Field, Harper & Collins, 2001

CHAPTER 13
DNA RESONANCE

Rupert Sheldrake, coined the term *morphic resonance* to describe the ability of living systems to inherit a collective memory from all previous forms of their kind.[1] The word 'morphic' is derived from the Greek word *morphē,* meaning 'form'. Reading Sheldrake's proposal, for the possibility of resonance between 'like lifeforms', my co-author, Peter Hewitt, and I realised there might be a link between morphic resonance in biological systems and the DNA molecule. We described this as *DNA resonance.*

In *The Vortex: Key to Future Science*,[2] Peter and I used the idea of DNA resonance to explain how the life field[3] discovered by Professor Harold Saxton Burr, might underlie the processes of differentiation in multicellular bodies and evolution. In biology, differentiation or morphology is considered to be controlled in chromosomes by genes switching on or off in a process which involves complex chemical reactions. But the research of Professor Burr suggests that these may be the consequence of a coordinating electric field matrix overlying the cell.

Is it possible that Burr discovered resonance between cells in the body and superenergy plasmic fields, overlaying and interpenetrating the cells? If so was the resonance occurring through the DNA in the nucleus of the cells? These were the questions Peter and I were asking. Peter was convinced vibrations in a field of superenergy could transmit a program for a cell via DNA much as broadcast electromagnetic fields can transmit a program via a tuned coil into a radio or television set.

The DNA molecule is a double helix coiled upon itself several times. This structure is reminiscent of the coil in a radio or television set that enables

1 Sheldrake R. A New Science of Life, Paladin Books, 1987

2 Ash D. & Hewitt P. The Vortex: Key to Future Science, Gateway Books, 1990

3 Burr. Blueprint for Immortality, Neville Spearman, 1972

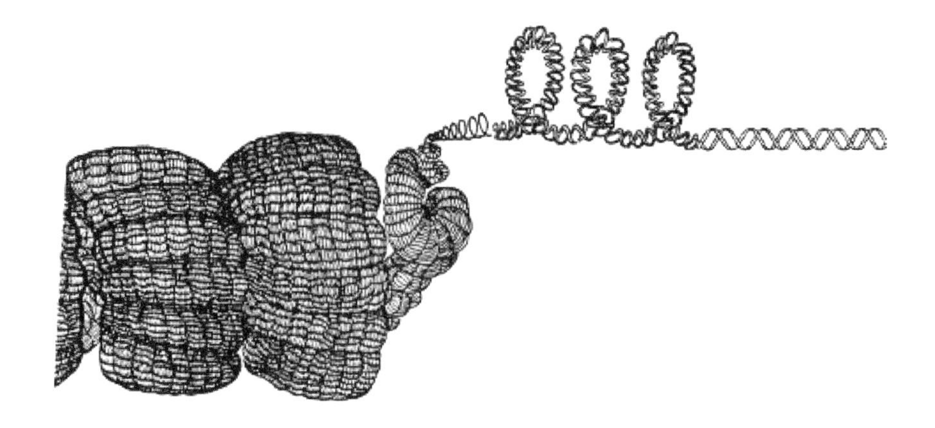

resonance with broadcast electromagnetic waves. Our thought was that the physical coil like structure of the DNA molecule might allow the life field, also known as a biofield, to resonate with DNA.

I was concerned as to how subtle superenergy could interact with dense physical DNA. Perhaps the fields of consciousness in the superphysical levels of reality, which correspond to spirit, might first resonate with the DNA in the hyperphysical soul body that ghosts the physical body. The program of information could then pass into the physical domain by resonance with the identical DNA in the cells of the physical body.

Resonance between matching DNA in the dense hyperphysical and physical levels of reality could enable the subtle fields of consciousness to penetrate the density of the physical world. The need for two stages of DNA resonance, first in the fourth dimensional DNA then in the third dimensional physical DNA, could be a contributing factor for why we have a body in the hyperphysical realm of reality as well as in the physical. I liken the fourth dimensional body to our need for soft undergarments before we don heavy overcoats to go out into the harsh conditions of the third dimensional world.

The resonance of the superphysical field of consciousness with the hyperphysical DNA, which then resonates with the physical DNA, could have similar effect to the amplification of a broadcast signal in the radio of TV set. It could increase the effective strength of the DNA resonance and so help the superphysical information penetrate the physical world.

The evidence of near-death experiences makes it clear that the mind and consciousness are carried by that part of us which separates from the physical body at death. Death may the moment when the DNA resonance between the hyper-physical and physical bodies ceases. The cells in the physical body do not die at the moment of death. Death is when the conscious, cognitive aspect of us leaves the physical body. That is what NDE research is showing.[4] Given the right conditions, the cells of the body can survive for up to 48 hours after death. That is why resuscitation is possible. When resuscitation occurs the conscious soul body is retrieved to the unconscious physical body. It snaps back into precise overlay with the physical body. That is the moment when death is reversed. The resuscitated person comes back to life. Resurrection has occurred. Resurrection may be simply the resumption of DNA resonance.

As Peter and I considered on the concept of DNA resonance, the interplay between the spirit, soul and body became clearer to us. Using the models of TV and radio we began to appreciate traditional spiritual ideas in a scientific context. It was Peter who suggested that a superphysical, spiritual, field of consciousness could broadcast a programme of information first to the hyperphysical soul-body, which then relayed that information into the physical body by DNA resonance. These ideas only made sense if the physical body is not the person but is a vehicle that a spiritual person occupies for the duration of a lifetime on the physical plane. It would be decades before the evidence of this would begin to appear in the scientific literature[4] but we were working on the science back then.

We were excited to realise DNA resonance might make sense of millennia of spiritual and religious teachings of body, soul and spirit and why religious people placed so much importance on the soul and so little on the physical body. As we considered the implications of resonance, through DNA as a means of transmitting information between the planes we realised we had a major breakthrough that could bridge the gap between science and spirituality. DNA resonance between superenergy and energy in the physical level of reality made scientific sense of spiritual phenomena. It literally brought the spirit into science.

Peter and I were working on a new paradigm where DNA Resonance could be the key not only to the origin and evolution of life on Earth but to the origins of humanity and the means whereby we, as spiritual entities, could operate as fully conscious beings on the physical plane. As my then wife Anna said to me one day, *"Resonance will be the keyword of the 21st Century."*

As well as Burr a number of other scientists have studied biofields. The British doctor Walter J. Kilner invented a series of goggles and filters through people could see biofield auras in detail. In his wake Harry Oldfield invented

4 . Parnia S. et al., Guidelines and standards for the study of death and recalled experiences of death - a multidisciplinary consensus statement and proposed future directions. Ann.NYAS, Vol 1511, Issue 1, P.5-21, 18-02-2022

Polycontrast Interference Photography (PIP) to register patterns of light, imperceptible to the human eye, that radiated from biological organisms. A Russian husband and wife team, Semyon and Valentina Kirlian, developed a way to photograph the biofield, using high frequency electric field resonance. All these researchers indicated the potential of using resonance to study higher dimensions of reality. They raise questions that could lead to new vistas of scientific enquiry. The research of Burr, especially, could be repeated to study the impact of electric fields on cellular differentiation. His work could do much to enhance our health and wellbeing.

DNA resonance points to the possibility of reciprocal information exchanges between different planes of reality. Directive information from the planes of superenergy could be downloaded via DNA resonance onto the physical plane but conversely experiential information could be uploaded from physical lifeforms back into the superphysical. This model has enormous scope for redefining our understanding of the interplay between body, soul and spirit.

When I was working on AWAKEN[5] with Matthew Newsome he drew my attention to the potential of DNA resonance in helping us understand biological feedback. He suggested that DNA resonance might shed new light on the *Akashic Records.* In Yogi philosophy all actions, thoughts and intents, words and emotions that have ever occurred are presumed to be recorded in a non-physical plane of reality called the *akashic*. Matthew thought that maybe information from the physical plane of reality could be uploaded through DNA resonance to some form of hyperphysical memory bank equivalent to a computer hard drive,.

The information exchanges between planes of energy through resonance might help explain how - as proposed by Sheldrake's morphic resonance - natural systems such as termite colonies, pigeons, orchid plants or even insulin molecules can inherit a collective memory from all previous organisms of their kind.[6]

An example of morphic resonance occurred in Southern England. In 1921, in Swaythling, near Southampton, someone reported a milk robbery to the police. A hole had been punched into the cap of a bottle of milk on the victim's doorstep and the top layer of cream had been stolen. The police suspected local youths but the culprit was eventually discovered to be a blue tit. Over the ensuing years sporadic reports of the early morning milk robberies popped up in various parts of the South East of England. Then in the 1950's something

---

5 Ash D & Newsome M, AWAKEN, Kima Global Publishing, 2018

6 Sheldrake, Rupert The presence of the past: Morphic resonance and the habits of nature, Icon Books, 2011

extraordinary happened. Milk bottles were being raided by blue tits, not only in England, but throughout Europe. Wherever milk was left in bottles on doorsteps it was liable to have the cream removed through a neat little hole in the cap. In the 1960's blue tits were following milk floats and it was hard to find an unpunctured bottle of milk on the doorstep, unless you woke up early and got to the bottles before the birds.

Research into the learning ability of blue tits indicated they didn't appear to learn this new feeding habit by observing each other. The rapid spread of the new behaviour, throughout England and across the channel into Europe, seemed to suggest there was some other mediating factor at play other than one blue tit mimicking another.

Matthew put forward a proposal to explain this example of non-local transmission of behavioral information between members of a species. His suggestion was that information from a single species individual in the physical domain could be uploaded into its hyperphysical brain by DNA resonance. The hyperphysical brain of the species individual, with the newly discovered information, then acted somewhat like a nerve cell in a brain. He suggested that the hyperphysical brains of all members of a species, in this case *Cyanistes caeruleus,* the blue tit, may be connected in a *hyperphysical species network,* somewhat like the nerve cells in a brain. Then just as one nerve cell in a brain can light up with information received from a sensory cell and pass it on to other nerve cells so in the hyperphysical species network information coming from one blue tit, via DNA resonance, could be passed onto other blue tit hyperphysical brains. From there it could be downloaded into their individual physical brains. In this way the learning from one species individual could be passed onto others in a non-physical manner.

I think Matthew's suggestion was brilliant. His idea that the hyperphysical brains of blue tits could be linked together in a hyperphysical network has endless ramifications. Imagine if the hyperphysical brains of our species were networked in the same way as the cells in our individual brains. It is an amazing thought that as a species maybe we could act like an individual brain.

Matthew's idea of *hyperphysical species networking* could apply to every species. It could account not only for the phenomenon Sheldrake described as *morphic resonance* but for the *hundredth monkey effect,*[7] which demonstrates that information learnt by one species individual isn't immediately passed onto all members of the species. It seems to require a minimum number of individuals to respond to a newly acquired pattern of behaviour before the whole species acquires it.

---

7 Keyes K. The Hundredth Monkey, C. DeVorss, 1984

This could be explained by brains responding to stimuli according to the mathematical law of exponential growth as expressed in the 'J' curve.

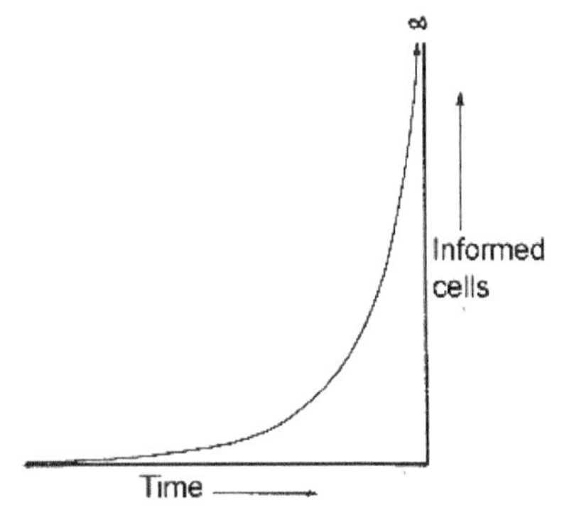

The first brain cell would inform its immediate neighbours and they would then pass the information onto their surrounding cells. Over each interval of time the rate of information transfer accelerates until a *tipping point* is reached. That is when the number of informed cells pass round the curve in the 'J' and the exponential growth occurs. This same law might apply to Matthew's *hyperphysical species networking.*

It makes sense that the passage of information through a *hyperphysical species network* would follow the pattern of exponential growth. That would explain why it took years for the information to pass from the blue tit population in and around Southampton to other blue tit populations in South East England, and then why suddenly, in the early 1950's practically every blue tit throughout Europe knew how to feed on milk by pecking through milk bottle tops., World War II may have delayed the exponential curve during the 1940's when there was a reduction in the number of milk bottles left on doorsteps. This might explain why the tipping point was delayed until the early 1950's when high volume milk deliveries were re-established after the war.

Hyperphysical species networks may be involved in the process of evolution. They may explain how evolutionary changes occur in numerous species individuals simultaneously and why an entire new species can suddenly appear. DNA resonance may be the means whereby information is uploaded and downloaded between species individuals and the species group acting as an interconnected whole. The physical plane may be where new opportunities are discovered by individual members of a species and where they are tested by natural selection but the hyperphysical plane may be where information and experiences of the group are processed and integrated within the species as whole.

Matthew's thinking suggests the physical plane may be where individuals gain experiences and learn lessons and the hyperphysical plane may be where lessons and experiences are integrated between lifetimes. Hyperphysical species networking could help to explain how the process of evolution might be informed and influenced by fields of consciousness. The theory of DNA resonance may help explain how the driving forces of evolution and learning through natural selection could operate between all three levels of reality: the physical, hyperphysical and superphysical. Advancement of a species could occur as follows. Creative innovation and design could happen in the fields of consciousness on the superphysical plane. These could then be tested on the physical plane through natural selection. Processing and integration of the tests and trials would then occur on the hyperphysical plane sandwiched in between. The information could be moving in fluid exchanges between the planes by means of DNA Resonance. The hyperphysical, acting as an Akashic archive, could hold a record of all species events in the hyperphysical species network, thus acting as a species memory.

If the superphysical is predominantly plasmic matter it could operate as a *plasmic field matrix* of electric fields. These superphysical plasmic fields of consciousness and intent could be endowed with animating intelligence and they might act as living templates capable of directing species and individual advancement. Rather than acting from a set pre-existing creator design, these templates could be subject to upgrades and modifications from experiences gained in an evolutionary process. Matthew's model offers a way intelligence could be involved in the evolutionary process. I like to think of the progress of life as a process of *creative evolution* made possible by DNA resonance.

Peter and Matthew made major contributions to my thinking around DNA resonance but one of my favourites was Peter's proposal that we take the pastoral god Pan as a metaphor for how evolution is creatively operating through DNA resonance. It was his suggestion to write in *The Vortex: Key to Future Science,*[2] that if Pan represented an intelligent living plasmic field intervening in the play of evolution, the pipes could symbolise DNA resonance. Peter wrote: *When Pan plays the same tune, the species remains the same. When Pan plays a different tune, the species alters.*

This thinking might be anathema to scientific orthodoxy embedded in the limiting philosophy of materialism but as Matthew said, *"People who still believe in materialism are like the proverbial frogs in a well that won't countenance the existence of the ocean".*

Graham Hancock asserted that the conditioning of scientific materialism has influenced educated people to reject the possibility of other realities that are more subtle and multidimensional than the apparent material world. In his foreword to Gregory Sams' *Sun of gOd,*[8] he wrote: *My personal experiences have opened my eyes to the possibility that the fundamental unit of reality is not matter but the spirit that animates and organizes it, and that there is no dichotomy between spirit and matter because these realms are so thoroughly*

8 Sams G. Sun of gOd, Red Wheel Weiser, 2009

intertwined and promiscuously interconnected. The problem rather is one of perception, that the logical, positivist, empiricist bias of Western science has so conditioned us to focus on gross matter, as though there is nothing else, that we have become impervious to the subtler fields of spirit that interpenetrate and surround it. With the help of others I have presented a superenergy model that is intended to help us appreciate in a scientific context what spirit is and how spirit and physical matter are thoroughly intertwined and promiscuously interconnected.

If DNA resonance can provide an explanation for how cells differentiate into tissues and organs maybe it could help us understand disease. Is it possible that weakening or disruption of a biofield could predispose an individual to disease? By stimulating the biofield through the medium of DNA resonance maybe it would be possible to strengthen or relax the fields within and around the body to help restore and maintain optimum health. The models of DNA resonance and the biofield, might help provide a scientific understanding of how subtle energy operates in alternative medicine.

A scientific account for alternative medicine and healing is much needed because in 1986, the British Medical Association published a report on alternative medicine[9] in which they concluded that while the therapies were effective they could not be endorsed because they were unscientific. There is no problem with alternative medical practice. Alternative medicine is very effective. The problem was and still is the unscientific materialistic paradigm that has been used as the benchmark to determine what is and what is not scientific. Now is the time to move beyond the limits of physicalism to consider evidence in support of superenergy. A major arena for consideration is alternative medicine.

9 BMA Board of Science, Complementary Medicine: New Approaches to Good Practice, Oxford University Press 1986

CHAPTER 14
ALTERNATIVE MEDICINE

Alternative medicine is effective. Yet within the limited theoretical framework of materialistic science it has proved difficult, if not impossible to explain how many alternative therapies work. However, if we expand our conceptual horizons to include the concepts of superenergy, the biofield and DNA resonance then many alternative therapies may become more understandable.

While many doctors and scientists dismiss alternative therapies as a placebo effect there is a growing serious interest in these modalities. There use comes under the regulation of Complementary and Alternative Medicines: CAMs [1][2][3] and many are now incorporated into the NHS.

The research of Harold Saxton Burr,[4] may help to shed some light on how alternative therapies work. His discovery of a life-field acting somewhat like an electric field could be a key because Burr's work suggests the life-field or biofield may be subject to the laws of electricity.

If superenergy is responsible for the biofield and the same laws of physics apply to physical energy and superenergy it should be possible to explain alternative medicine in terms of physics. Acupuncture especially can be understood in terms of electric field effects.

In physics I was taught that electrostatic fields are evenly distributed over spheres but if there is a spike on the sphere then a charge will accumulate on the spike. When that occurs a potential difference builds up between the

---

1 Sixth Report – Complementary and Alternative Medicine; House of Lords Select Committee on Science and Technology, November 2000

2 Lewith G.T. et al Complementary Medicine: Evidence base, competence to practice and regulation, Clin. Med., 2003 May- June (3): 235-40

3 What Is Complementary and Alternative Medicine? National Centre for Complementary and Alternative Medicine

4 Burr H.S. Blueprint for Immortality, Neville Spearman, 1972

spike and the body of the sphere. Electrical energy is then able to flow down the *potential* gradient between the spike and the sphere. These simple laws of electrostatics could help in understanding acupuncture.

If a superenergy biofield overshadows a physical body the electrostatic principles would suggest that potential gradients could develop between the body and extremities on it such as ears, fingers, toes, nose and the limbs. And indeed Burr measured voltage differences between digits, limbs and the head and body.

Burr's research seemed to indicate a flow of energy down the potential gradients between the head, the body and the extremities. Could it be that a resonance occurs between the hyperphysical body and the physical body, which is picked up as an electric field effect? Is the potential gradient in this field indicative of a flow of superenergy in the life-field? Could this correspond to *Chi*, the life-energy that the Chinese recognise to flow through the body, which is capable of maintaining or restoring health and vitality?

The Chinese discovered flow lines of Chi between every organ in the body and the limbs, which terminated in the hands and feet. They also discovered that Chi flows on the head to the ears and nose. It seems the flow of Chi follows electrostatic principles.

The Chinese called the energy flow lines, or channels of Chi, *meridians.* They found that if they stimulated or sedated the flow of Chi in the meridians they could restore the balance of Chi - the life force energy - in the organs. They did this by inserting needles into specific points on the meridians related to the organs. This system of healing in China came to be known as *acupuncture.* [5]

5 Geng J., Selecting the Right Acupoints: Handbook on Acupuncture Therapy New World Press, China 1995

An intermediary for the action of acupuncture, through the resonant flow of Chi, may be the *interstitium.* The interstitium is a network of fluid compartments just under the top layer of the skin which connects to tissue layers lining the organs, gut, blood vessels and muscles. The interconnecting interstitium compartments are supported by a mesh of strong and flexible proteins capable of generating electric currents. The interstitium also produces and conducts the flow of *plasma* lymph, which is capable of conducting electricity between the skin and other tissues in the body. In acupuncture the electric activity of the interstitium may be stimulated or sedated to increase or reduce the flow of currents of life force between the skin and other major organs.

Another therapy based on the principles of acupuncture is *acupressure.* In acupressure the Chi, or life-energy, is balanced by massaging acupressure points on the skin that are related to specific organs.

Meridians - the life energy flow lines - related to every organ in the body terminate at the feet. Applying acupressure to the terminal point of a meridian on the soles of the feet appears to stimulate healing in its associated organ.

This system of life energy balancing is called *reflexology.*

In 1986 when the British Medical Association reported on alternative medicine they said they could not support it because it was unscientific. They cited reflexology as a case in point.[6]

A modality is considered to be unscientific if it cannot be accounted for in the materialistic frame of understanding. Superenergy provides a new scientific paradigm in which if alternative medical practices can be accepted as scientific. This is especially important for homeopathy which was the most widely accepted medical practice before the pharmaceutical industry was established. Homeopathy is safe and effective with minimal side effects but despite this, the practice of homeopathy is considered to be fraud by many doctors and professors and sceptics even though innumerable people testified to its efficacy.

When Professor Jacques Benveniste, a medical researcher in France, performed experiments that appeared to validate homeopathy he had a veritable visit from the inquisition. John Maddox, editor of *Nature*, accompanied by the Australian magician and leading sceptic, James Randi, descended on his laboratory outside Paris and pulling his records apart they turned his lab upside down. A campaign of disinformation followed to discredit Professor Benveniste. As a result he lost his job and his reputation was ruined.

Homeopathy was established over 200 years ago by a German Dr Hahnemann[7] on the principle of *like cures like.* Hahnemann found that if a substance causing a particular set of symptoms was greatly diluted then the diluted form of it could cure the symptoms. For example, arsenic will cause severe stomach pains but Hahnemann discovered that an extreme dilution of arsenic could cure stomach pain. Hahnemann

Samuel Hahnemann (1755–1843)

6 BMA Board of Science, Complementary Medicine: New Approaches to Good Practice, Oxford University Press 1986

7 Gupta H. and Hahnemann S. Medicine for the Wise: Hahnemann's Philosophy of Diseases, Medicines and Cures, Create Space Independent Publishing Platform, 2014

found that the more he diluted his remedies the more effective they became. This was because he *potentised* his homeopathic remedies by percussing them in a diluting procedure of mixing and pounding peculiar to homeopathy.

I found the superenergy model could provide a simple account for homeopathy. In the process of potentising a homeopathic remedy the energy field of the substance, which caused the symptoms could generate an energy field of the substrate - a neutral material such as sugar or water - in which it is diluted. The energy signature of the substance would then act like a *photographic positive,* imprinted on the energy field of the sugar or water substrate by the pounding or energetic mixing in the potentising process. This produced an opposite energy impression in the substrate, acting as the equivalent of a *photographic negative.*

With increased dilution the substance generating a positive impression would be decreased. With ever more potentising the negative energy impression would be increased. In this process the concentration of the symptom causing substance would be reduced while the quantity of symptom curing substrate would be increased. This explains why homeopathic remedies become more effective as they are diluted and potentised.

Benveniste discovered that homeopathic remedies acted through intent as well as the potentising process. He discovered this when he placed the name of a remedy on a slip of paper under a glass of water the water became empowered with the therapeutic properties of the named remedy. This suggested a direct action on the water by superenergy fields of consciousness as though the practitioner had declared for the remedy with a *prayer-like* intention. A review of healing might help us appreciate how this could work.

CHAPTER 15
HEALING

My father, Dr Michael Ash (1916-1991) was a Harley Street physician. He was one of the earliest medical practitioners of acupuncture in the UK and was also a gifted healer.[1] I recall how he hovered his hands over a patient, a few centimetres above the skin. He moved his hands to stimulate a flow of healing energy. He explained to me, when I was a boy, that the science of healing was similar to moving a magnet to simulate a flow of electricity.

I now believe my father was stimulating a flow of superenergy into the biofield of the patient to revitalise their cells, realigning them in the natural hyper-physical body template. My father believed a potentially unlimited reservoir of energy could be tapped by a healer and directed by intent to the life-field of the patient. I realised from my vortex physics that separation between the healer and the patient in the third dimension would not affect the superenergy connection in the fourth dimension. This was how I explained my father's ability to heal people at a distance.

Distant healing has been practiced since ancient times and there are remarkable healing stories in the Gospels where Jesus was able to regenerate people instantly. One story of his ability to heal someone at a distance, recorded in the Bible, is in the Gospel of Matthew (8:5-13). A Roman centurion asked Jesus to heal his servant with palsy. Jesus said immediately he would come and heal the servant but the Roman replied that he was not worthy to have Jesus in

1 Ash M., Health, Radiation and Healing, Darton, Longman & Todd, 1963.

his home. He said that Jesus had only to say the word and his servant would be healed. Jesus was amazed at the man's faith and told him to go home because his faith was rewarded. When the centurion returned to his home he found his servant was recovered. When other servants in the house told him when the miraculous recovery occurred he realised it was when he had been with Jesus. This action of Jesus healing by intent could help us appreciate Benveniste's remarkable demonstration that intending the name of a homeopathic remedy was as effective as administering the remedy itself.

My father practiced distant healing on a number of patients every morning before breakfast. I recall him working on a child in Texas. I was seventeen and we were living in Sussex. The child's father said the child would stir and cry every night at about the time my father would focus on him and send healing from England. My dad asked the child's father to record the time when the child awoke. Allowing for the time difference between Texas and England they found the child stirred when my father sent the healing intent. The child recovered from leukaemia but sadly died from an infection introduced by a blood transfusion.

My father was a close friend of Harry Edwards who was also an exceptionally gifted healer.[2] Harry had a healing centre at Shere in the Surrey Hills. My father took me there and introduced me to Harry and he told him I was going to develop a science to explain healing. He reminded me on a number of occasions that I was destined to provide a scientific account for healing.

I was able to tell my father that the spiritual energy flowing through him and healers like him could be coming from worlds of superenergy and that spiritual worlds were levels of quantum reality existing beyond the speed of light. I said to him because the speed of light is slower than the speed of energy in the spiritual worlds, our world would be part of a superenergy spiritual world and spiritual beings in the spiritual world could see us and work with us even if we couldn't see them. All they needed to be effective was our faith in them and their ability to help us. Calling on a remedy could be as effective as administering a remedy if we stepped aside and allowed them to do the work. I explained that psychic phenomena was mostly the result of healers, mediums and psychics working with supporters in the superenergy spiritual dimensions. I also suggested that asking angels to help us find things was a practice that might improve our connection with the superphysical worlds.

My father liked my account as he enjoyed mixing with psychics and spiritualists as much as he enjoyed mingling with scientists and medical practitioners. He also had an interest in physics relating to radioactivity and geo-pathology.

2 Edwards H., Harry Edwards: Thirty Years a Spiritual Healer, Jenkins 1968

From his medical practice and geopathology research in the West Country my father discovered the danger of an increased risk of cancer from living above uranium. He was the first doctor to raise the concern. Also, in the 1950's, he was the first medical doctor to raise the alarm that cigarette smoking causes lung cancer. He was ahead of his time.

My dad was a polymath physician and a physicist, a geologist, a dowser and a faith healer. As well as introducing me to nuclear physics, when I was seven, he taught me and my brothers and sisters how to heal when we were little. I grew up in a world where science and medicine were mixed with psychic studies and healing.

He perceived healing in terms of the operation of electric and magnetic forces. Working with my hands as a healer, in the way my father taught me, I could feel the force he had worked with as a tingling in my fingers, as could my brothers and sisters who he also taught to heal.

My father said superenergy was the *odic force.* He said the odic force fitted my description of a field of superenergy overlaying the physical body. The odic force was a name given by Baron Carl von Reichenbach, in 1845, to the biofield, which was confirmed experimentally a century later by Harold Saxton Burr. My father introduced me to Burr's work. He was interested in Burr's research into the life-field and he had a high impedance voltmeter for repeating Burr's experiments. Burr's description of an energy template controlling the differentiation of cells helped me develop my ideas of superenergy fields ghosting the physical body.

Father used dowsing to demonstrate that the superenergy biofield extends a few inches beyond the body. He told me the healthier the body the further the biofield extended out and that healing energy would flow from his healing hands into his patients like water flowing downhill or electricity flowing down a voltage gradient. That fitted neatly with my principle of energy speed subsets.

My father endorsed the way I explained the transference of healing energy from the spiritual planes to the physical plane in terms of resonance between the biofield and the physical body. He taught me that the biofield could act as a template for the physical body because it was not affected by aging or disease or wear and tear incurred by a physical body. I was able to use my vortex theory to explain to him that the *body template* maintained its integrity because it was not in physical space or time and therefore was not subject to physical influences. I explained because the physical world was part of the superphysical dimension it was subject to spiritual influences. This, I said, was clearly demonstrated by spiritual healing.

Harry Edwards described himself as a *spiritual healer*. Through his intent superphysical energy from aligned fields of consciousness was directed into the biofields of his patients. His faith and the faith of his patient enabled superenergy to flow from higher dimensions into them to enact regenerative healing in their body much as Jesus did.

Never charging for his services, Harry's faith and purity of intent enabled him to channel superenergy from the highest levels of consciousness into the superenergy biofields of afflicted people. Harry was stimulating the biofield body template of his patients, sometimes seen as an *aura*, to assist them in the regeneration of their physical bodies. My father and Harry said they never did anything remarkable as healers. They just directed a flow of energy and supported a patient's faith in their own natural healing ability.

Psychics and healers sometimes see the biofield as an aura of light surrounding the body. An exceptionally bright aura of light, seen around the head corresponds to a halo.

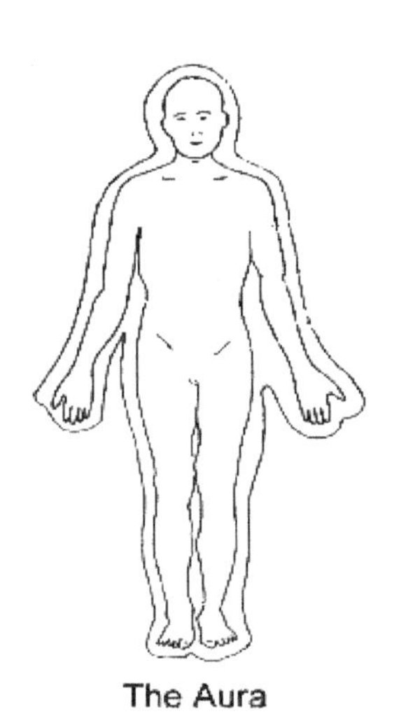

The Aura

Awareness of the aura, by seeing or feeling, occurs when people have the faculty of psychic perception. This is usually inherited. Perception of the aura by psychic people in ancient times could explain why Plato asserted the existence of a psyche that carries the animating force of life into a body at conception and departs it at death.

The descriptions of nature spirits and elementals could have been the way people in the past perceived the biofields associated with plants. The innate intelligence in biofield auras, associated with trees and plants, could account for why sensitive people sometimes speak to their plants and hug trees. Developing a scientific account for what many people experience as an energetic connection with the natural world could help us all to respect, acknowledge and communicate with the plant kingdom. This can help us become more sensitive to the environment and its fragile ecology. We can become more aware by working with plants. Spending time in nature and talking with plants, even house plants, can help us be at peace in turbulent times and feel more serenity in our lives.

Psychics and healers sometimes perceive colour in the human aura. These they relate to emotions and the health and overall disposition of the patient.

The colour frequencies in the aura can be used by healers to diagnose disease.

Having the properties of an electric field the aura can be photographed using equipment that involves high voltage and high frequency electric fields that resonate with the biofield to give images of a corona of colour lights.

Aura images were first captured on a photographic plate, in 1939, by a Russian electrical engineer Semyon Kirlian and his wife, Valentina. The electro-photography they developed to capture images of the aura came to be known as Kirlian photography. The vivid colours in the aura are depicted in *aura photographs*, which is now widely available through Kirlian photography. Anything that conducts electricity can give an aura photograph but Kirlian photography works on non-conductors picking up, by resonance, the bioelectric effects of the aura. Auras around living things are colourful and it is the colour in the auras, also seen by psychics, that gives them significance.

The Kirlian's received a Russian patent in 1949 and first published their work in 1958 but because of the Cold War it was 1970 before their work reached the West. While Kirlian photography was endorsed by the Russian Academy of Science in 1966,[3] in the West it was treated with disdain and dismissed as a hoax.[4] The problem we are faced with is not a lack of evidence but the resistance in universities to pursuing the research necessary to gather the evidence. Thelma Moss was a professor at the University of California at Los Angeles (UCLA) when she opened a scientific laboratory to research Kirlian photography in the early 1970's. The lab was unfunded and run by volunteers but its line of research into the resonance effects between living organisms and high frequency, high voltage electric fields was deemed to be outside the bounds of legitimate science and therefore was not tolerated by UCLA. The lab was closed down by the university in 1979.[5]

---

3 Juravlev, A. E., Living Luminescence and Kirlian effect, USSR Academy of Science, 1966

4 Stein G. Encyclopedia of Hoaxes. Gale Group, 1993

5 Greene S, UCLA lab researched parapsychology in the '70s News, A Closer Look. UCLA Daily Bruin, Oct. 27, 2010

CHAPTER 16
THE HOLY GRAIL

Without doubt the most renowned healer in recorded history was *Yesua the Nazarene* anglicised as Jesus of Nazareth. After a lifetime of healing the sick, ministering to the poor and preaching the Word of God, he was rewarded for his efforts by crucifixion. There is a legend that when Jesus was dying on the cross his blood was collected in a cup which was the most cherished relic in Christendom. But the relic was never found. The reason for this is that that story of the Holy Grail was misunderstood. It was not a cup of the physical blood of Jesus that held the power of salvation, it was another cup of his blood. Let me explain.

The Gospel of John begins: *In the beginning was the Word and the Word was with God and the Word was God. The same was in the beginning with God. All things were made by him; and without him was not anything made that was made.* (John 1:1-3) *And the Word was made flesh and dwelt amongst us...* (John 1:14)

Jesus was the flesh that incarnated the Word of God. When Jesus spoke it was the Word of God that spoke through him. When he performed a miracle, it was under instruction of the Word of God that the miracle occurred. The Word incarnated as Jesus but was not restricted to Jesus. The Word of God could overlay any body it chose. The physical body of Jesus happened to be the biological vehicle that was chosen to enable the Word of God to walk and talk amongst men and women in Palestine two thousand years ago. Through the physical form of Jesus of Nazareth, the Word could heal, teach and perform miracles. It was the Word of God incarnate in Jesus doing these things not the physical Jesus. When Jesus said, *"I am the light of the world"* it was the Word of God speaking through him, not Jesus the man speaking as normal people would converse.

Many people believe that Jesus Christ will return. The Word of God has returned but not as the man Jesus. The word Christ means the *anointed one* and the Word of God can choose to anoint and appoint anyone at any time to best fulfil God's purpose and the Word of God is not bound by the expectations or religious institutions of man. The physical form of Jesus was a garment that enabled the supernatural Word from a higher dimension in the Universe to act in the lower physical world. The body of Jesus was nothing more than a biological garment for the Word of God. The Word of God is free to choose any physical vehicle and it may not be human flesh and blood. As it is the Word chose to return in the form of bread and wine, which we could consume so the Word could dwell in us just as the Word had dwelt in the body and blood of Jesus the Christ.

When Jesus was crucified, the physical garment of God was effectively hung up on a peg. The Word of God had already left it. This was made clear in the Bible which records Jesus lamenting on his cross, *"My, God, my God, why hast thou forsaken me."* (Mark 15:34) The Word had moved into a new form of body and blood that would enable the Word of God that had come in the form of Jesus Christ to return throughout the world on a daily basis and act for the spiritual wellbeing of humanity, for generations to come, after Jesus was gone.

On the night before he suffered horrific death by torture on the cross, Jesus held a Passover meal with the disciples he had hand-picked to carry forward the real *Holy Grail* that was about to be established: *AND as they did eat, Jesus took bread, and blessed, and broke it, and gave to them, and said, "Take eat: this is my body." And he took the cup, and when he had given thanks, he gave it to them and they all drank of it. And he said unto them, "This is my blood for the new testament, which is shed for many."* (Mark 14:22-24)

The many are Christians who believe that Jesus established a covenant with his disciples that whenever they or their anointed successors uttered those words of consecration over bread and wine, the bread and wine would become the new form of matter that the Word of God would inhabit. Jesus had completed his mission on Earth. On the night he was betrayed the Word left his body of flesh and blood for a new body of bread and wine that people could eat and drink so they would take the Word of God directly into their hearts. That was the Holy Grail of redemption. That is how humanity would be saved. Once we received the Word of God in our hearts, the Word could assist us in our attempt to live according to the way we were taught through the teaching and example of Jesus. By receiving the Word as bread and wine we allowed the Word to help us. This didn't mean we could just rely on the Word to save us without any effort on our part. We still had to do the work of love, forgiveness and prayer in our daily lives but the Word of God that came in the consecrated bread and wine, known as the Eucharist, could give us the strength to persist in our practice of faith despite internal resistance and external opposition.

Many people find it hard to believe that the Eucharist is the equivalent of the body and the blood of Jesus Christ but the science of superenergy can help us appreciate the mystery. When consecrated as the Eucharist, bread and wine can act in same way as the body and blood of Jesus. While the Word cannot speak to us through the vehicle of bread and wine it can carry the Word of God into us. On receiving the consecrated bread and wine, the Word of God can overlay and resonate with our hearts, according to the principles of superenergy subsets and coincidence, as it overlayed and resonated with the body and blood of Jesus. The inter-relationship of superenergy and physical energy can help us understand how the Word of God, as a superenergy field of consciousness, can be as present in the consecrated bread and wine as surely as it was in Jesus. It only requires faith for this miracle to occur. When we consume the consecrated food we intend by that act to bring the Word of God, which Jesus had incarnated, into resonance with our soul. One has to admit it is sheer genius.

Jesus Christ was the gifted operator of a science of superenergy which enabled him to perform healings and miracles. He was not unique. In his day Apollonius of Tyana had similar powers and in our own time a miracle man has appeared with similar gifts. By studying him we can get a better understanding of the physics of miracles.

CHAPTER 17
A MAN OF MIRACLES

Born Sathya Narayana Raju (1926-2011), Sathya Sai Baba, as he become known, was a world famous miracle worker, Indian guru and philanthropist who resided in Andhra Pradesh, India.

The display of miracles throughout Sai Baba's life led to an enormous following in the West as well as in India. He was famous for his ability to manifest valuable jewellery for his devotees.[1] These included rings, pendants and expensive watches. Those fortunate enough to receive one of his lavish gifts treasure them for the miraculous manifestation of faith they represented.

Many recipients of Sathya Sai Baba's gifts had them authenticated by jewellers and the gold was always real gold and the gem stones genuine. Sai Baba may have died but his rings and watches live on, worn on the fingers and wrists of numerous singularly blessed individuals.

Sceptics said that Sai Baba was an exceptionally clever magician. However, it is very unusual for magicians to give away their props, especially expensive items of jewellery. When asked how he performed his miracles Sai Baba explained that the rings, watches and ornaments already existed in his *Sai stores.* All he had to do was visualize them in his mind and they would appear in his hand.

1 Dabholkar H. Shri Sai Satcharitra: The Wonderful Life and Teachings of Shirdi Sai Baba, Enlightenment Press (2016)

In India the ability to perform miracles is called a *siddhi*. The siddhi of materialisation was acquired by Sathya quite suddenly when he was fourteen. In March 1940 he lost consciousness for twenty four hours after being stung by a scorpion. On recovery he had a spontaneous spiritual awakening. In the weeks that followed the teenager started a mission as an advanced spiritual teacher and miracle worker.

When the fourteen year old awoke from a twenty-four hour coma his family found his behaviour very strange. He alternately laughed and cried uncontrollably. Then he quoted long passages of Sanskrit philosophy and poetry that he had never learnt and he described in vivid detail distant places he had never visited. For two months his family called on the help of an expert in devilry to exorcise the demon they believed to have taken over their son but to no avail.

Then, on the morning of May 23rd 1940, the boy performed a miracle. In front of his family and some people from his village he materialised flowers and candy out of nowhere. The attempts at exorcism were immediately abandoned. Someone in the amazed gathering called out, "Who are you?" Sathya replied calmly but firmly: "I am Sai Baba". He explained he had been reincarnated through the prayers of an Indian sage, one of India's most revered modern saints, Sai Baba of Shirdi,[1] who had died in 1918.

Shirdi Sai Baba, as he was then known, had preached the importance of self-realisation and criticized love of perishable things. His teachings concentrated on a moral code of love, forgiveness, helping others and practicing inner peace, charity, contentment and devotion to God and guru.

Shirdi Sai Baba condemned distinction based on religion or caste. He was neither Muslim nor Hindu. Religious distinctions were of no consequence to him so his teachings combined elements of both Islam and Hinduism.

The villagers and Sathya's family struggled to accept the possibility that Shirdi Sai Baba had come again. Gathering around him they demanded he show them a sign. With a quick flip of his hand he threw a bunch of jasmine flowers onto the floor. In the language of the village they spelt out, *Sai Baba*.

I believe Sathya Narayana Raju was killed by the scorpion sting. After his soul left the dying body the soul of Sai Baba of Shirdi resurrected it and moved in. In modern parlance this would be called a *walk-in.* It was Sai Baba of Shirdi who woke up from the coma in the fourteen-year-old body not Sathya Narayana Raju. Attaching the boy's first name to his own, Sai Baba continued his ministry in India, that had been terminated by his death in 1918. Returning in a new body, twenty-two years after he had died, he came with more spiritual power than he had when he was alive as Shirdi Sai Baba.

Sai Baba in the hyperphysical world would have come into resonance with the physical body when the soul of Sathya Narayana Raju ceased to resonate with it. After Sai Baba healed it from the effects of the scorpion sting he resumed life where he left off with all the knowledge and experience from his previous incarnation. This would explain how the fourteen-year-old was able to quote long passages of Sanskrit philosophy and poetry and describe in vivid detail distant places unknown to the young Sathya Narayana Raju.

The *walk-in* process is very similar to a near-death experience except that when the soul separates from the dying physical body it doesn't return. Instead, after resuscitation a different soul wakes up in the body. I call the walk-in process *soul-swapping.*

Sai Baba's newfound ministry accelerated at a rapid pace. By 1944, *Sathya Sai Baba,* as he now called himself, had attracted a considerable following and his devotees built a Mandir (Hindu temple) near his home village of Puttaparthi in Andhra Pradesh. In 1948 they started to build an ashram for him, which was completed in 1950. By 1954 Sai Baba had established a small, free general hospital in the village Puttaparthi, which was the beginning of his lifelong philanthropy. The fame Sathya Sai Baba garnered from his miracles and his ability to heal spread far and wide, across India. Then, in the 1960s, he became known in the West.

A professor of psychology Erlendur Haraldsson, studied Sathya Sai Baba closely. He reported his assessment of Sai Baba in a scientific paper[2] and wrote a book about his experiences.[3]

Sathya Sai Baba stood out in the bright orange robes he always wore and his jet black Afro hairstyle. There was no mistaking him. Meeting him in person was an unforgettable experience. When he performed his miracles he would reach up and pluck an item out of the air so that sceptics could not accuse him of using sleight of hand to slip jewellery out of the sleeves of his long orange robes.

---

2 Haraldsson, E. Appearance and disappearance of objects on the presence of Sathya Sai Baba. J.Am.Soc.Psy.Res., Jan. 1977.

3 Haraldsson E. Modern Miracles: The Story of Sathya Sai Baba: A Modern Day Prophet, White Crow Books, 2013

A compelling testimonial comes from a Californian visitor to Sathya Sai Baba's Ashram. The American wanted to photograph the guru but found the film in his camera had run out and he had omitted to bring a spare roll. Sai Baba reached up and plucked a roll of film for the camera out of thin air and handed it to the Californian. The man took photographs and on his return to the USA he went into his local photo store to get the roll of film developed. When he returned to collect the pictures the owner of the store remarked on the photographs. He said the person in the photos had called into his store a week before to buy a roll of film. He remarked on it because he was an unusual customer dressed in bright orange robes with an Afro hairstyle. When asked when it was the store owner recalled without difficulty. Sathya Sai Baba had walked into his store and bought the roll of film when the Californian was standing with him in India. Apart from one trip to Kenya and Uganda in 1968, Sathya Sai Baba never left India and yet there are numerous reports of him appearing in countries outside of India.

A firsthand account of Sai Baba's materializations that throws a light on how he performed his miracles comes from the report of an interview with Sathya Sai Baba by the famous American spiritual teacher, Ram Dass: *When I was there, as I was sitting at his feet and he was sitting on a chair, he said to me, "'Here Ram Dass, I'll give you something." and I said, 'No Babaji, I don't want anything.' "No, no, let me give you something." He held out his hand, and I knew he did things like this, manifest small things like bracelets, watches, small things like that.*

As a social scientist, responsible to the West, my eyes were going to watch his hand closely, I wasn't going to blink. As I watched, a bluish light formed on the top of his hand, a flickering light, and it became more and more solid, and then it became a little medallion. It was a little circle a star on it with a little gold image of himself, Sathya Sai Baba. He gave it to me, it was definitely man-made, it did not have an astral quality to it at all.

Later I asked a Swami there, "How does he do that?" And he said, "Well, he doesn't make those; he just moves them from his warehouse with his mind. And you can just imagine his warehouse, full of these little medallions, and if you were in the warehouse, they'd be disappearing from the shelves, literally."

This report by Ram Dass helps to explain how miracles performed by Sathya Sai Baba could occur. The objects manifested by Sai Baba existed in the physical domain prior to their apport or miraculous apparition. This is clear from an inventory taken after Sai Baba died in 2011. He had gold and silver ornaments, mainly in the form of jewellery and watches, warehoused in his personal quarters of the ashram. The medallions had been bought by the ashram but the watches and jewellery were given to him over many decades by his Indian devotees. It is common practice in India to lavish gifts on a guru.

Sathya Sai Baba lived up to his message - *Love all, Serve all* - by giving generously throughout his life. He built schools and hospitals and provided humanitarian support to the poor of India. But giving poor people expensive jewellery would not have done them any good. Instead Sai Baba used his expensive gifts to help spiritual seekers and visitors from the West to believe in him. The Westerners, who came to India to see Sai Baba were not poor in material things but many of them were spiritually impoverished. They needed proof of supernatural and spiritual dimensions to support their flagging faith and to encourage them to follow the principles Sai Baba wanted them to live by: *Truth, Peace, Righteousness* and *Love.*

When Sai Baba came back into the world, on March 8th 1940, he brought with him remarkable spiritual gifts, which caused him to become one of the most famous Indian Gurus of all time. His powers were similar to those of Jesus Christ. Two thousand years before Sai Baba, Jesus performed healings and miracles to help people have faith which caused him to become the most renowned spiritual teacher in the West. As with Jesus, Sai Baba's healings and miracles supported the spread of his all-important message of universal love. The downside for both of them was that their followers thought they were God. This problem can be overcome by using the concept of superenergy to explain manifestations and miracles in a scientific rather than a religious way.

My suggestion is that miracle workers like Jesus and Sai Baba could have been assigned operators of a superenergy technology on the hyperphysical plane who worked in tandem with them on the physical plane. Sai Baba could have been working in association with someone in the spiritual fourth dimension in order to help people place a renewed emphasis on the spiritual life. Sai Baba certainly achieved that objective. As a spiritual teacher and miracle worker, as Jesus did before him, Sai Baba opened the hearts and minds of millions of people – including my in-laws – reminding them that there is more to life than material wellbeing.

My premise was that Sathya Sai Baba would have been familiar with the gifts of watches and jewellery he had been given, which he kept in his Sai stores. During an interview with a Western seeker Sai Baba visualised one of these objects. In so doing he used *telepathy* to inform his hyper-physical operator of the item he wanted. The operator, aware of Sai Baba's thoughts, then directed a beam of superenergy onto the intended object on a shelf inside the Sai stores. Once the item was in the superenergy of the beam the energy in its subatomic vortices resonated with the super-energy. In the process it became superenergy and no longer able to remain in physical space-time it evaporated into hyperspace. That was my explanation for how it vanished off the shelf. As Sai Baba stretched out his hand the watch or piece of jewellery in hyperspace was translocated out of the Sai stores into his hand. This would have been achieved by the operator shifting the beam, with the item suspended in it, just above Sai Baba's outstretched hand. The hyperphysical operator, in the fourth dimension, then switched off the beam. The energy in every quantum vortex and wave, in the matter of the watch or piece of jewellery, immediately reverted back to the speed of light. That caused the item to *condense* back into physical space and drop neatly onto Sai Baba's hand. The item of jewellery or watch that a moment before had been in Sai Baba's store was now in his hand. It would have been moved through hyperspace not physical space. While in hyperspace the object was invisible to the visitor. All they would see was the object dropping out of thin air into Sai Baba's outstretched hand before it was presented to them. This was not magic or sleight of hand, nor an illusion. It was the application of a superenergy technology that can be understood in terms of the superenergy physics.

Immediately before being switched off if some of the superenergy in the beam above Sai Baba's hand condensed into physical energy it could have appeared as physical light. That explains the momentary appearance of the flickering blue light Ram Dass saw as his medallion appeared.

Sathya Sai Baba's *bilocation* to California can be understood in terms of *duplicate body replication.* In my theory this could have been achieved by streaming superenergy through Sai Baba's physical body template in India to produce a duplicate physical body that was evaporating into hyperspace as it formed and then condensing back into physical space in America to appear near the photo store. This enabled Sai Baba to walk into the store in California and buy a roll of film while he was standing in the interview room at his ashram in India. After walking out of the store the duplicate body could have been evaporated back into hyperspace and dissolved into superenergy leaving the roll of film to be condensed into Sai Baba's outstretched hand in India. The replica body could be described as a *holographic insert.* But why did Sathya Sai Baba go to all the trouble of duplicating his body in order to acquire a roll of film when he could have evaporated it off the shelf in the shop. The answer is simple. That would have been shop-lifting. Sai Baba had to materialise in America in order to purchase the roll of film.

In her book *We The Arcturians*[4] Norma Milanovich describes an Arcturian

4 Milanovich N. We the Arcturians, Athena Publications., 1990

technology that enables a living body to be scanned, molecule for molecule and cell for cell so it can be backed up on a data base. A replica body can then be reproduced - much like 3D printing - as a holographic insert to appear anywhere in the Universe. This may have been the way Sai Baba's bilocation was achieved. However they happen, bilocations and miracles can be explained in terms of superenergy resonance.

CHAPTER 18
SUPERENERGY RESONANCE

Progress is made in science by proposing a hypothesis or formulating a law and then testing it against a body of evidence. Near-death experiences and the miracles of Sathya Sai Baba provide evidence to support the idea that realms of superenergy exist beyond the speed of light. I contend that the only difference between the world of physical energy and the worlds of superenergy is the intrinsic speed of energy in the photons of light and quantum vortices of matter that make them up. So long as the laws of physics in all the worlds are the same then bodies should be able to move freely between worlds simply by changing their intrinsic speed of energy. This could occur by a process called *superenergy resonance.*

My explanation for the miracles performed by Sathya Sai Baba is that physical objects were being overlaid by a beam of superenergy operated from a higher dimension. When this occurred they underwent *resonance* with the superenergy in the beam which caused their intrinsic speed of energy to accelerate until it matched the speed of energy in the beam. In doing so the object ceased to be a body of physical energy and became a body of superenergy. It vanished out of the physical third dimensional world and was suspended in the superenergy beam in the hyperphysical fourth dimension for translocation. Then when the beam was switched off the resonance ceased and the object immediately appeared in a new location in the third dimension. This procedure was used to move objects from one place to another in Sai Baba's ashram by shifting in and out of space and time rather than through it. Sai Baba also visited the USA without visibly leaving his ashram. Did that happen much as Captain Kirk was beamed by Scotty off a planet into the star ship *Enterprise*? While only a story, Startrek presented a graphic depiction of superenergy resonance, which in essence is the speed of energy in a physical body increasing until it is in equilibrium with the speed of superenergy in the beam.

Thinking about the *"Beam me up Scotty"* story and the miracles of Sai Baba, it struck me that superenergy resonance might be nothing more nor less than a system of energy coming into equilibrium with its energy environment. Energy equilibrium occurs when you place a hot object in a cold room and it cools down or if you place a cold object in a hot room and it heats up. To rationalise superenergy resonance in terms of energy systems establishing and maintaining equilibrium I proposed two cosmic *laws of motion.*

Cosmic Law of Motion I:	***A body of matter will maintain its intrinsic speed of energy unless it is caused to change.***
Cosmic Law of Motion II:	***If the cause of change is removed the intrinsic speed of energy in the body will revert to its original value.***

The first cosmic law of motion is exemplified by a body of matter undergoing a change of state from physical energy to superenergy due to resonance in a superenergy beam. The second cosmic law of motion is demonstrated by reversion of the body to its original energy state when the beam is turned off. The cosmic laws of motion can be depicted by a spring.

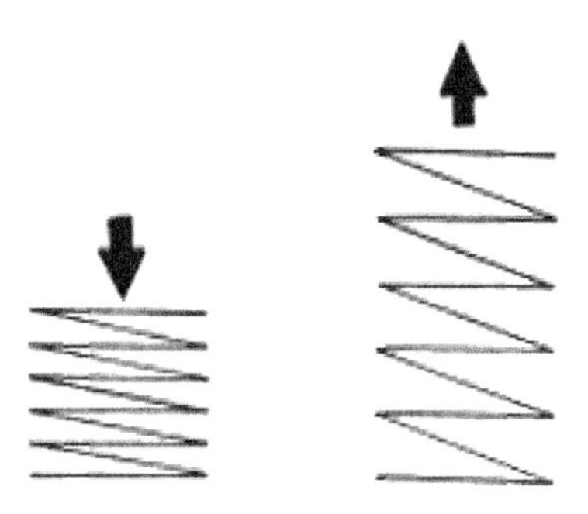

In line with the cosmic laws of motion I realised if a body of superenergy were to enter physical space its intrinsic speed of energy could drop into equilibrium with the lower speed of energy in the physical space. That caused me to wonder if the speed of energy might differ in different regions of the Universe. I was in Australia on a lecture tour in the early 90's thinking about the story of the fall of grace at the time of Adam and Eve in terms of a fall in speed of energy. I realised the fall might have occurred as a result of the solar system entering a region of space where the speed of light is depressed below a normal value. This could be caused by an increase in density much like the way the speed of light is depressed when it enters water or glass. I called the hypothetical regions of depressed energy zones of *density.*

It struck me that maybe our solar system was once hyper-physical or even superphysical. If it underwent a depression in its intrinsic speed of energy as a result of entering a dense region space then the cosmic laws suggested if ever it left that region of space it would automatically recover its original energy state. At the time my lectures included the subject of human and planetary ascension. Leaving a zone of density made perfect sense in terms of what I was teaching. Years later I read a prophecy by Peter Deunov that lent credibility to my suppositions. I have included the prophecy at the end of this book because Peter Deunov says our solar system has been traversing a contaminated region of space for a very long time that has caused a period of metaphoric darkness described in India as *Kali Yuga*. He also says we are about to leave this region of contradictions for a clearer region of space. When we do we will *ascend* into a time of metaphoric light or enlightenment and universal peace and harmony.

The cosmic laws of motion also throw light on why we might grow a superenergy body in parallel with our physical body. The laws suggest while we could make temporary transitions between the worlds of energy and superenergy we cannot remain in the worlds unless we have a body to live in that is made of the energy of that world. They suggest that to live in the physical third dimension we need a physical body. To live in the hyperphysical fourth dimension we would need a hyperphysical body. To live in the superphysical fifth dimension we would need a superphysical body. The cosmic laws of motion make sense of how we could live in different dimensions and transition between them. This is the physics of ascension.

The ability to translocate a body while it is in hyperspace is sometimes called *pinging.* Superenergy resonance holds a promise of pinging that would render obsolete cars, trains, aeroplanes and passports. National boundaries would be a joke and no one would ever be incarcerated again. I visited a prison in Sidney in 1992 to give a talk on this subject to prisoners. Needless to say they loved it. Afterwards I was told by prison staff that an Indian gentleman had been in the prison for some contravention of the law. One day he announced he was going home as it was his daughter's birthday. They told him he wasn't free to come and go as he chose but being a yogi he smiled sweetly, shook them by the hand and pinged out of prison. Very quickly they collected him from his family home and reincarcerated him but after a couple of weeks he decided it was time to ping again. After that the prison system gave up on him. They couldn't hold him so they let him go. The ping technique he used was a bit more sensational than just vanishing. He walked through the prison walls.

To explain superenergy resonance I use as metaphors the terms *evaporation* and *condensation* for ease of understanding as they denote changes of state. In superenergy resonance the changes of state are in the intrinsic speeds of energy between physical-energy and superenergy. In the superenergy resonance between physical and hyperphysical dimensions a body evaporates into a hyperphysical state and condenses back into physical state. Conversely a body of hyperphysical matter could be condensed into the physical and then evaporated back into the hyperphysical. The latter process could account for *UAPs* (Unidentified Anomulous Phenomena) and *apparitions.*

Apparitions sometimes occur in a condensation of superenergy called ectoplasm. When they evaporate they disappear in the ectoplasm which melts away like mist. This is why it is sometimes called *ghost mist.* The luminous ectoplasm often shrouds the apparition which has led to people donning a sheet when they want to dress up as a ghost. In times gone by this was called magic or miraculous but with the physics of superenergy it is possible to explain such supernatural phenomena in terms of an understanding of resonance.

Ectoplasm is generated by a form of superenergy resonance, which involves the transition of energy rather than bodies of matter between the worlds. I call this *vortex resonance.* In vortex resonance superenergy can be condensed into physical matter and serve as a medium through which a holographic insert can then be condensed into physicality. Mists of ectoplasm form as superenergy is condensed through a physical body template. When Sai Baba appeared in the USA superenergy was streamed through his physical body template and evaporated for transition as a duplicate physical body to be relocated by condensation in California.

In vortex resonance superenergy could be streaming through the matrix of molecules, atoms and subatomic vortices in any body of physical matter acting as a template. Imagine the superenergy jetting through the atomic matrix in the manner of a laser beam. As the superenergy jet condenses through the physical body template into physical energy it could replicate the physical matter in the manner that Sai Baba's body was replicated. You could think of this somewhat like 3D printing. Jesus could have used this form of vortex resonance to feed the five thousand. Deployment of vortex resonance could also explain his healing miracles. Superenergy streaming through the physical body template could stimulate instant regeneration at a cellular level.

In another form of vortex resonance the super-energy does not condense into energy through a beam. Instead a vortex in the physical dimension resonates with a vortex in the hyperphysical and superenergy passes from the hyperphysical vortex into the physical vortex and condenses into physical energy. This form of vortex resonance could be a source of boundless free, sustainable energy.

CHAPTER 19
PSYCHIC SURGERY

Psychic surgery is a supernormal procedure through which many thousands of people have experienced dramatic healing. Materialists decry psychic surgery as a hoax. They accuse psychic surgeons of fakery and say the patients have been duped. This disparaging attitude stems from the inadequacy of materialistic science to explain paranormal phenomena and the trend of educated people, embedded in materialism, to deny, ignore or explain away the supernatural. The downside of this attitude is it blocks research into superenergy technologies that could benefit humanity and help solve some of the problems that beset us.

The strange phenomenon of psychic surgery first came to public attention in 1959 when Ron Ormond published *Into the Strange Unknown.*[1] He called the practice 'fourth dimensional surgery,' and wrote: *We still don't know what to think; but we have motion pictures to show it wasn't the work of any normal magician, and could very well be just what the Filipinos said it was — a miracle of God performed by a fourth dimensional surgeon.*

In the psychic surgery, first discovered in the Philippines, the psychic surgeons are often illiterate. Certainly they don't have any medical training. They usually start by running a hand over the skin of the patient. As they do so a hole appears as though a scalpel has been used, only no scalpel is ever used and there was no pain or bleeding. The psychic surgeon then plunges his hand into the hole, sometimes with a kitchen knife that may be blunt and never with an anaesthetic, and performs an operation.

The psychic surgeon might remove a cancerous tumour, a blood clot or some other obstruction. That is when some bleeding occurs. At the end of the operation the psychic surgeon then passes his hand over the hole and as he does so it disappears. The skin is left intact, as it was before the operation.

---

1 Ormond R., Into the Strange Unknown, Esoteric Publications., 1959

There are no stitches, no pain and no scar.

John of God, working in Brazil, became the most famous psychic surgeon of modern times[2] [3] Starting out as a farmer, he gave up three days a week to practice psychic surgery, free of charge. Then he had to work at it full time as thousands of people come to him from all over the world. One of them was the mother of my younger son.

Psychic surgery can be explained in terms of superenergy resonance. If a laser like beam of superenergy is applied to a specific part of a patient's body the energy in every subatomic vortex and quantum wave in that region of the patient's body would *resonate* with the superenergy in the beam. In the resonance the intrinsic speed of energy, in the tissues now in the beam, would accelerate to match the speed of the superenergy in the beam. The atoms, molecules and cells in the beamed flesh would then undergo a change in state from physical matter to hyperphysical matter. In line with the first cosmic law of motion, the tissues in the beam would *evaporate* out of physical space. While the superenergy resonance beam is on, the psychic surgeon could perform his operation in the hole that has appeared before him. After the operation, in line with the second cosmic law of motion, when the superenergy resonance beam was switched off the body tissues that were temporarily suspended in hyperspace would immediately revert to the physical state. That is how the flesh, *condensing* back from hyperphysical matter to physical matter, could suddenly reappear again.

The Filipinos said that a fourth dimensional surgeon was operating. This suggests that a hyper-physical surgeon was directing the superenergy resonance technology. It is possible this operator was a surgeon before he died. He could now be working from the hyperphysical fourth dimension in tandem with the psychic surgeon in the physical third dimension.

The hyperphysical surgeon would have had no choice but to work with an illiterate person because educated people are taught to question and dismiss the supernatural. In doing so they block it.

2 Cumming H., John of God: The Brazilian Healer Who's Touched the Lives of Millions, Atria Books, 2007

3 Meliana M. John of God: A Guide to Your Healing Journey with Spirit Doctors Beyond the Veil, Blue Leopard Press, 2014

The illiterate psychic surgeon indicated the area to be operated on by a sweeping motion of his hand - much as a nurse indicates to a surgeon where to cut by drawing a line on the skin of a patient with a marker pen. In the case of psychic surgery the flesh would be parted not by a scalpel in the hands of a physical surgeon but a superenergy beam directed by an unseen hyperphysical surgeon. In this extraordinary procedure the physical tissue that became invisible did not go anywhere and was not disturbed by a surgical incision. It just passed beyond physical perception and interaction. There would have been no change in the molecular or cellular structure of the tissues during the superenergy resonance procedure. It would have become a zone of superenergy. Blood flowing through the vessels within the operational zone would have been temporarily suspended in hyperspace. In its passage through the zone of superenergy the blood would have undergone a change of state to hyperphysical blood. On leaving the zone of superenergy surgery the patient's blood would immediately revert to physical blood.

On completion of the operation, when the superenergy resonance beam is switched off, the tissues would reappear in physical space. As there was no physical incision there would be no need for healing and there would be no scars. This is a straightforward account for psychic surgery but any attempt to explain such paranormal phenomena or understand miracles is impeded by both religion and science. Religious people proclaim miracles as acts of God while scientists discredit them as hoaxes and people like John of God are dismissed as charlatans. The physics of superenergy holds the promise of breaking this pattern by providing a straightforward account in physics for the supernatural. Rescued from suspicion and superstition psychic phenomena could become the next frontier of scientific discovery. One phenomenon that calls for research is ectoplasm because of its extraordinary behaviour and the way it is involved in the phenomenon of ghosts, the condensation of deceased people in real bodies onto the physical plane.

CHAPTER 20
ECTOPLASM

In her autobiography[1] Doris Stokes, gave a very graphic description of an ectoplasm apparition:

Silence fell, Helen Duncan concentrated deeply and then appeared to go into a trance. This was quite routine and by now I had seen it happen several times, yet there was something electric in the air. Something strange and tense that I'd never noticed before.

As we watched a thin silvery mist began to creep from the medium's nostrils and her middle, yet she remained motionless in her chair as if she were asleep.

"Ectoplasm," someone whispered behind me. Gradually the flow increased, until the mist was pouring from the medium and a wispy cloud hung in the air in front of her. Then like fog stirred by a gentle breeze, it began to change shape, flowing and swirling, building up in places, melting away in places.

Before our eyes the outline of a woman was being carved in mist. Hair and features began to sharpen and refine. A small nose built up on the face, then a high brow, lips and chin, until finally the swirling stopped and she stood before us, a perfect likeness of a young girl in silvery white - and she was beautiful.

My mouth dropped open and I couldn't tear my eyes from this vision. I was seeing it, yet I couldn't believe it. Dimly I was aware that the woman next to me had gasped and clasped her hands to her mouth, but before I could register the significance of this the girl began to move.

The audience watched, riveted as she drifted across the room and stopped right in front of my neighbour.

"I've come to talk to you mother," said the medium in a light, pretty voice quite different from the one she'd used earlier. The girl spoke to her mother, for several minutes, explaining that she still visited the family and knew what was going on

1 Stokes D. F. Voices In My Ear, Futura Publications, 1980

and listed a few personal details as proof.

Then unexpectedly, she turned to me. "Would you like to touch my hand?" she asked.

Dumbly I brushed the slim, pale fingers held out to me, and then in astonishment took the whole hand. It was warm! I don't know what I had expected. Something damp, cold and unsubstantial I suppose - but this was incredible. I'd touched a warm, living hand.

Suspiciously I glanced at the medium but she was still slumped in her chair. It was impossible. It must be a fake and yet how could she have done it? Nonplussed I sank back and stared at the girl, quite speechless.

She smiled as if she could read my thoughts, then she raised her arm and out of the air, a rose appeared in her fingers. Gently she placed it on my neighbour's lap.

"Happy Christmas mother" she said then slowly moved back and began to shrink, getting smaller and smaller, fainter and fainter until she disappeared through the floor.

No-one stirred. We all sat motionless as if hypnotized. The only sound was the woman next to me quietly sobbing. In her hand a deep red rose, still beaded with dew - in December. It was only later I discovered that Helen Duncan was one of the greatest materialisation mediums who ever lived.

In a cloud of luminous plasma called ectoplasm a deceased girl coalesced into physicality. She was solid and warm, a real living person. She then melted away like early morning mist. The living body that appeared in ectoplasm, was recognised as the daughter who had recently died by her mother in the audience. The fact that Doris felt the hand and found it was solid and warm indicated she was not having a vision. Everyone else in the room saw the girl too so physical light was reflecting off her body. Doris felt her with her fingers so it was not an illusion. The girl was real and alive. She was formed of physical matter, as are you and I. She was a human being recognised by her mum. She had died and was now standing in the room. The electrons in the physical atoms of her body repelled the physical electrons in Doris's hand. That was how Doris felt her as a solid body. The ectoplasm was also visible which suggests that it was physical matter too. Then there was the rose. That was plucked out of the air by the girl and given to her mother. The girl vanished but the deep red rose remained on her weeping mother's lap; complete with a drop of dew.

The way the girl plucked the red rose out of nowhere was reminiscent of the way Sai Baba plucked jewelry out of thin air. If the medium was working with a hyperphysical operator using superenergy resonance, then in line with the first cosmic law of motion the operator could have evaporated a rose from a garden or nursery anywhere on Earth then translocated it through hyperspace into the room. If the beam was then switched off, in line with the second cosmic law of motion, the rose would then have condensed back into physical space into the outstretched hand of the recently deceased girl. The rose was physical and so was stable in the physical domain.

To explain the apparition of the girl who had died, superenergy could have been streamed through the physical body template of the medium so that it could condense as matter on the physical plane. Once it formed the replica body of the deceased girl could coalesce out of it, through her hyperphysical body template, and appear as a holographic insert onto the physical plane. The ectoplasm, or ghost mist enabled the holographic insert to coalesce as a real, warm, solid, living body in physical matter. To witness this was a very rare and special opportunity and quite remarkable.

While only a story the fable of *Aladdin* and the lamp, from *The Arabian Nights,* has a similar tone. It may have been based on ectoplasm apparitions that occurred in ancient times. When Aladdin rubbed the lamp smoke came out of it in which a genii appeared. (The word genii is derived from the Persian word for *spirit.*) The smoke could have been ectoplasm in which the genii condensed into physicality. If that were so and Aladdin was a real boy he would have experienced the apparition of the genii much as Doris Stokes witnessed the condensation of the deceased girl.

Aladdin's genii fits the image of an operator of super-energy resonance. The difference here was that the operator condensed from ectoplasm into physicality as a holographic insert to receive his orders. The story goes on to depict him moving things and people, including Aladdin, in and out of physical space-time. That fits with superenergy resonance.

The immensely popular story of Aladdin and his genii, which has fired the imaginations of generations for centuries, is an outstanding example of how a fable in ancient folklore may convey important elements of fact. Ancient traditions of magic, depicted in stories and legends like Aladdin, can help us appreciate how much truth there may be in folklore.

When human beliefs and values are relegated to stories, they are not especially debased because stories are often repositories of truth. In *Braiding Sweetgrass,*[2] professor R. W. Kimmerer emphasizes the importance of stories in the preservation of culture and traditional teachings in indigenous societies. In our own society stories weave fact with fiction. Historical novels are an outstanding

---

2 . Kimmerer, R.W., Braiding Sweetgrass, Penguin, 2013

example. They can be invaluable teaching modalities. The story of Aladdin may be just that.

A feature of the Aladdin story that is of particular interest is the lamp. In the story of Aladdin a lamp acted as an *anchor point* through which ectoplasm flowed for the condensation of the genii. Helen Duncan acted as an anchor point through which ectoplasm flowed for the condensation of the girl.

If ectoplasm is a condensation of superenergy into physical energy, through the medium of a physical template, which allows the apparition of a hyper-physical body to coalesce in the physical state, ectoplasm could be serving an important purpose. If the cosmic laws of motion are correct then objects formed of physical matter would stable in physical space but unstable in hyperspace. Holographic inserts from the hyperphysical plane, such as ghosts or the genii or the girl, would be unstable on the physical plane. The condensation medium of ectoplasm appears to be necessary to stabilise them and help maintain their physical presence.

The stabilising effect of luminous ectoplasm was evident in series of Marian apparitions that occurred in recent times in Egypt. The anchor point through which the ectoplasm condensed into physicality was a church roof. These apparitions were viewed by thousands of people who came to Cairo to see them and millions more when they were shown on television worldwide. I was transfixed by the sight of one of these apparitions featured on the early evening news, when I was living in Bexhill-on-sea, Sussex in 1968.

From 1968 to 1970, Mary appeared on the dome of a Coptic church in Zeitoun, Cairo, Egypt.[3] Night after night, for well over a year, the outline of a lady with a halo appeared in luminous ectoplasm which pouring from her as she bowed and moved about. The ectoplasm flowed over the roof and dome of the church, lighting up the night sky with its brilliance.

The Marian apparitions in Zeitoun fitted the description of an ectoplasm generated holographic insert. They were unprecedented and undeniably genuine. According to local legend, Zeitoun was the village that harbored Mary and her family when they were refugees in Egypt. These apparitions helped me verify the cosmic laws of motion and added credence to my theory of superenergy. They also helped me to believe in miracles and in the truth of Marian apparitions and the supernatural. After seeing the apparition of Mary on television I determined to develop a science of the supernatural to dispel the superstition of scientific materialism that demotes everything miraculous to the mundane. Years later my endeavours were recognised and rewarded.

3 http://www.zeitun-eg.org/zeitoun1.htm

In December 1989 I was blessed with an apparition of Mary in the church of Medjugorje in Bosnia-Herzegovina. When she appeared words came into my head to ask her to introduce me to her son. She disappeared and he appeared immediately, standing at a distance on a hill in a white robe flowing in the wind. I could feel the wind blowing on my face. It was hot and my cheek facing into the wind was stung by particles of blown sand.

I was in a state of ecstasy. My then wife Hillary, sitting next to me, attested to that. She said that I was frozen and my head was thrown back and tears were streaming down my face. I can only imagine I was transported to the middle East in a body to have felt the hot wind and sand stinging me on the face. I wonder if I was taken back into a previous lifetime to meet Jesus when he was alive in Palestine? I know I was close to him in a past life and if that were so maybe I was taken back to meet him in a situation when I was with him physically.

If our soul bodies from every life live on in the hyperphysical dimension it should be possible for us to step out of time into any one of them and then re-enter the physical situation when we were alive in that body. In effect we would be stepping back in time. This would make sense of the Akashic records. If everything that has ever been still exists then it should be in our ability to revisit our past by slipping in and out of time. This is my interpretation of the experience I had of a hot wind and sand stinging my face.

Imagine time is a river we float on as we progress through our many lifetimes. At any point we can swim to the bank and nip back to a previous point in the river to dive in and enjoy the experience of being there for a bit before returning along the bank to rejoin the river where we left it. That would be analogous to our ability to move in and out of space-time and revisit ourselves in a previous life. Imagine if we could relive our past lives to learn more from them.

During my apparition experience a light came from Jesus' heart which filled me with liquid light that was also love and intense joy. It moved through my body and ended up in my hands which were frozen and shaking with power. I asked Hillary to open doors and lead me back to our room in the village. At my request she placed a manuscript of an earlier version of this book [4] into my hands and the power went whoomph into the book. I had taken the manuscript with me to Medjugorje hoping Mary would bless it in some way. She must have heard my prayer but instead of doing it herself she very kindly arranged for her son to come and consecrate my writing with the light from his Sacred Heart.

Ectoplasm and anchor points did not occur in my apparition because my consciousness was taken into another situation when I was in the trance. This is more common in Marian apparitions. A physical apparition, like at Zeitoun, is very rare. It was necessary in the first apparition at Fatima in Portugal, in 1917. When Mary first appeared to the three children she was luminous which was suggestive of ectoplasm and she anchored on a Holm oak tree. When Jehovah first appeared to Moses in the wilderness he anchored on a bush. The appearance of the bush on fire without burning may have been the impression Moses had of luminous ectoplasm. A common pattern in apparitions is for a holographic insert in ectoplasm to appear for the first time to individuals as a means of introduction. Subsequently they are taken into ecstasy, as I was, to meet the divine in a higher dimension. If the individual is in a church they can be taken into ecstasy immediately. The appearance of a preliminary, holographic-insert ectoplasmic apparition is more common outdoors.

There are stories of Apollonius of Tyana and Jesus Christ vanishing from threatening situations. According to the Roman records Apollonius disappeared from a trial instigated by the tyrannical Roman Emperor Domitian who had determined to kill him. He then reappeared several hundred miles away. Jesus disappeared from a crowd set on stoning him death and was then seen walking at a distance. These events could have occurred through superenergy resonance. To remove themselves from danger Jesus and Apollonius could have used telepathy to instruct their hyperphysical operators to evaporate them into hyperspace and translocate them to a safe distance from the threat. The superenergy beam would then have been switched off enabling them to condense back into physical space in the safe location. In line with the cosmic laws of motion these translocations of Apollonius and Jesus did not require ectoplasm because they were already in the physical dimension. When Sai Baba bilocated to America he did it without ectoplasm. Because he had a

4 Ash D. & Hewitt P. The Vortex: Key to Future Science, Gateway Books, 1990

physical body on the physical plane its replica would have been physically stable.

Jesus and Apollonius suffered persecution as did Helen Duncan. She suffered imprisonment in the 20th century on an accusation of witchcraft. Even today her persecution goes on. Hoaxers have been active on the internet planting malicious deceptions on pages pertaining to paranormal phenomena. These are evident on the Wikipedia site where Helen Duncan has been discredited by slanderous accusations and the insertion of fake pictures suggesting that she used dolls and muslin to deceive people. There is no substantial evidence to support the malicious accusations falsified against her.

The time has come to apply science to the supernatural instead of against it. There is a need for the scientific method to lead us out of the morass associated with miracles where miracle workers are either maligned or mistakenly worshiped as God. People who perform miracles are not God, they are humans gifted with supernormal powers.

People who decry the paranormal as a deception and discount it as a hoax have a right to voice their opinions. For me it doesn't matter because with the science of the superenergy. I know it is only a matter of time before they will be more widely recognised as the equivalent of flat earth thinkers, but I make no excuse for people who hoax in the name of science. Anyone who falsifies evidence in the name of science, because it threatens their blind faith in the philosophy of materialism, makes a mockery of science and impedes its progress.

CHAPTER 21
CROP FORMATIONS

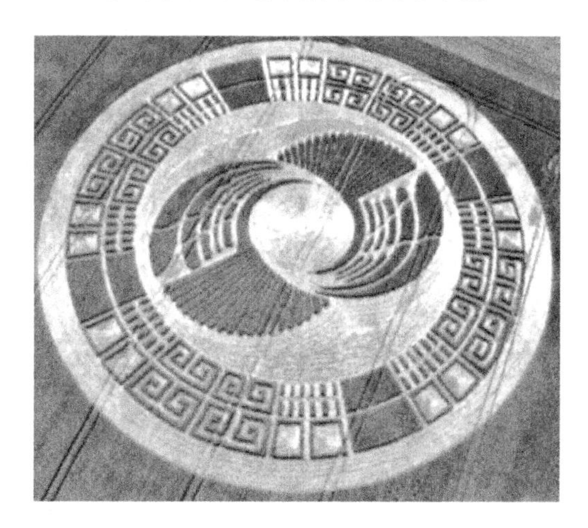

One of the problems with research into the paranormal is that when reports of new phenomena appear in the media and begin to arouse public interest, as occurred with crop circles, the malicious deception of *hoaxing* occurs to discredit the psychic phenomenon. This is then followed by confessions of hoaxers claiming to be responsible and to have been unmasked. The media then report without nuance that the entire phenomenon is a hoax. Despite the number of examples that can't be disproved the majority of people believe what they hear or read in the media and lose interest.

The paranormal is not a hoax. It is the hoax that is a hoax. The exposure of hoaxing in the media is the hoax. Hoaxing is ploy to discredit the paranormal. It is a deception intended to prevent people from questioning science or religion.

Awkward facts that challenge the generally accepted view of reality are either ignored or discredited and most people disbelieve in things that have been discredited. It is very easy to manipulate public opinion in the controlled media and destroy the credibility of anything that conflicts with what the religious, scientific and political establishment want us to believe.

The chants of 'hoax' or 'conspiracy theory' against any challenges to the status quo reveals the objective of *the establishment* to prop up what they want us to believe. This is why what is false appears as true and what is true appears as false. Meanwhile, the news of really important things remains underground. Paul Hawken in his book *Blessed Unrest* commented:[1] *I wrote this book primarily to discover what I don't know. Part of what I learned concerns an older quiescent history that is reemerging, what poet Gary Snyder calls the great underground, a current of humanity that dates back to the Paleolithic. Its lineage can be traced back to healers, priestesses, philosophers, monks, rabbis, poets and artists who speak for the planet, for other species, for interdependence, a life that courses under and through and around empires.*

Most scientists refuse to investigate phenomena that doesn't fit with what is currently held to be scientific. If they were true scientists, they would treat all phenomena with impartiality. They would be interested in expanding the boundaries of science. To do otherwise reduces science to a sham.

The hoaxing hypothesis supports the consensus worldview based on unquestioned belief in what is taught in church at school and at university. People cannot pride themselves as genuine sceptic thinkers if they never question consensus beliefs in science and accept blindly that hoaxing can account for practically all paranormal phenomena, especially crop formations.

Until the claims of crop-circle hoaxing appeared, it was difficult to establish a serious case against hoaxing. However, the story of 'Doug & Dave,' stumbling out of the pub, night after night, to create enormous crop formations in numerous different locations over short summer nights is totally untenable. If crop formations are a hoax there must have been an army of well organised hoaxers at work, night on night, throughout every summer season for decades to account for all the crop circles that have appeared.

The notion of an army of tipsy hoaxers cavorted through corn fields in dead of night, tramping on rope-tied boards to produce flawless patterns of a geometric magnitude never seen before on the face of the Earth, beggars belief. It is beyond the bounds of credulity that hoaxers would be that devoted to maintain the level of activity required to account for all of the crop formations reported, filmed and photographed since the 1980s.

Parties of people could not work the fields, night after night, in good weather and bad without ever being detected. They could never create the vast, complex, perfect geometric formations that have appeared, year after year in the few hours of summer darkness, without ever leaving tracks or getting caught. People park vehicles and make a noise; they use torches which send out lights

1 Hawken P. Blessed Unrest, Viking, 2007

that can be seen. People leave muddy footprints and make a mess. They could never get away with the decades of unremitting trespass required to make the high volume of formations; especially when irate farmers are on the lookout.

The precision and complexity of the vast crop formations, the sheer number of them and the way they suddenly appear in daylight hours as well as at night, explodes the hoax hypothesis. The crop formation phenomenon has unmasked hoaxing. Hoaxing is not the play of pranksters. It is a word thrown at anything mysterious or inexplicable which catches public attention in an ongoing campaign of disinformation intended to discourage serious interest in supernatural phenomena.

It takes only one genuine crop formation to disprove the hoax hypothesis that all crop formations are hoaxed. It only requires one genuine, inexplicable crop formation to raise the question of the origin of the remarkable unexplained pictograms that appear in crops, worldwide, year after year. While there are a vast number of unexplained formations very few scientists have investigated them. One who did conduct meticulous research into the phenomenon of crop formations was biophysicist, William C. Levengood. In 1994, the *Journal of Plant Physiology* published a paper by Levengood that reported unusual anatomical anomalies found only in plants taken from crop formations.[2]

When examined under the microscope, Levengood established the formation of plant tissue is markedly different when taken from unexplained formations compared to samples taken from hoaxed formations. In hoaxed formations plant stems are buckled or snapped as the crop has been clearly trampled while in the unexplained formations the stalks look as though they have been carefully laid down into a bent position with no sign of breakage.

For decades W. C. Levengood scrutinised crop formations and documented physiological changes associated with them, not only with the plants in the formations but in the soil under them. A formation that appeared in Cherhill, Wiltshire, England, in August 1993 was investigated by Dr Levengood. His findings were so remarkable that they were published in *The Journal of Scientific Exploration* in 1995.[3] The journal reported: *...the unusual discovery of a natural iron "glaze" composed of fused particles of meteoritic origin, concentrated entirely within a crop formation in England, appearing shortly after the intense Perseid meteor shower in August 1993. Abnormalities in seedling growth was also consistent with the unusual responses of seeds taken from numerous crop formations...Presence of meteoric material adhering to both soil*

2 Levengood W.C, Anatomical anomalies in crop formation plants, Physiol. Plant 92, 356, 1994

3 . Levengood W.C. Semi-Molten Meteoric Iron Associated with a Crop Formation, J. Sci. Exploration, Vol.9, No.2.pp 191-199, 1995

and plant tissues, casts considerable doubt on this being an artificially prepared or "hoaxed" formation.

W.C. Levengood discovered microscopic fused particles of iron unique to crop formations, much like grapeshot, in the soil under genuine formations. He found the concentration of fused particles of iron diminished in proportion to distance from the centre of the pattern. Observing these melted particles in the soil Levengood concluded that: *...very large amounts of energy radiation must have been involved in the formations, with the intensity falling off with distance from the centre of the formation. The science points to a complex, chaotic, thermodynamic energy field of enormous intensity, with components acting unpredictably and independently.*[4]

Levengood's fastidious observations are incongruent with the hoax hypothesis, especially considering the discovery of node changes in the standing plants as well as the bent plants in the unexplained formations. The evidence defies the simplistic view of hoaxers flattening the crop with ropes and boards. The evidence reveals the deployment of a thermodynamic technology to bring about the changes in the plants and underlying soil observed in the many genuine formations investigated over years by Levengood.

To quote author, John Mitchell, from the foreword to Crop Circles: *The Hidden Form*[5] by Nick Kollerstrom: *After some twenty years of crop circle research no one yet has any idea of what is going on. Every season new and better designs appear in the cornfields. They are amazingly subtle and beautiful. Nothing in the world of art today has anything like their quality...In the early days it seemed plausible that the circles were caused by freak whirlwinds or some other weather effect. That idea became impossible after 1990 when the first elaborately-designed 'pictograms' appeared. These, obviously, were products of intelligent minds. So the theorists were divided. UFO enthusiasts believed the intelligent source to be extra-terrestrial, while most other people took the down-to-earth view that it was all a hoax.*

The 'hoax' theory implies that unknown teams of skilled and dedicated artists are secretly at work during the summer nights, stamping or raking out large scale patterns and leaving no evidence behind. That seems the only rational solution. Yet there are so many difficulties to this explanation that experienced researchers are skeptical. No one ever detects or catches sight of these supposed circle-makers, or their cars or equipment. Certain 'hotspot' fields are watched during the summer crop formation season, yet circles suddenly appear in them

4 Levengood W.C & Talbot N.P., Dispersion of energies in worldwide crop formations, Physiol. Plant 105, pp 615-624, 1999

5 Kollerstrom N., Crop Circles: The Hidden Form, Wessex Books, 2002

over night, and nothing has ever been seen or heard.

Then there is the problem of how these large complicated patterns could possibly be completed in the few hours of summer darkness, never left unfinished and never showing any visible error. Copyists have been commissioned to make their own circles, legally, in daylight and with no time limit. But none of these has ever managed to come up with anything to match the quality of the great unclaimed masterpieces that appear spontaneously during the British summer months every year.

The observations of scientists like Levengood dispel the notion that all crop formations are hoaxes. His study of formations points to complex energy fields of enormous intensity being responsible for genuine formations. In recent decades they have increased in size, complexity, magnificence and dispersal in England and abroad, yet no effort has been made by the mainstream scientific community to investigate them. The simpler style of early formations could be duplicated by hoaxers but this changed in the 1990's when increasingly complex pictograms appeared that were too intricate to be mimicked. This development, ignored by the mainstream media, coincided with the blanket dismissal of all crop circles as hoaxes by the scientific community.

Science is supposed to be impartial but when it comes to crop formations the lack of scientific impartiality is fully exposed. The scientific establishment now stands shoulder to shoulder with the religious establishment in terms of denouncing evidence in defense of faith. Fortunately, there are still a few real scientists amongst minions of shams. They have the integrity to stand up and speak out about the urgent need to investigate crop formations. When Professor Gerald Hawkins spoke on crop formations in 1997, at a meeting of the American Astronomical Society in Washington D.C., he received a standing ovation. Hawkins had been speaking about what he called the intellectual profile of the unknown artists. He said, "*...the mechanics of how the crop patterns are formed is a mystery but the intellectual profile behind it all has turned out to be an even greater mystery.*" [6]

Crop formations provide evidence in support of superenergy and the existence of other worlds. If the other worlds exist beyond the speed of light crop circles could validate the principles of energy speed subsets and coincidence.

Genuine crop formations can be explained if we allow for the existence of intelligent beings with advanced technology in worlds we cannot see but which permeate our own. The superenergy hypothesis can explain why we only ever

6 Hawkins, G., From Euclid to Ptolemy in English Crop Circles, Bull. Am. Astronm. Soc., 29, p 1263, 1997

see pictograms in the fields. We never see the intelligent beings behind them. But they could be aware of us and certainly would be aware in our world if they use our fields of crops as canvasses for the execution of their exquisite works of art without causing them harm.

The execution of crop formations could be likened to the art work of the elusive British graffiti artist, Banksy. Banksy prepares stencils for one of his works of art in a studio. He then ventures out under the cover of darkness to a section of a wall or building facade and in a short space of time, when nobody is watching, he sprays aerosol paint through his stencils and then he promptly disappears leaving a masterpiece of art on the wall to be discovered the following morning.[7]

I like to think that artists in the hyperphysical level of reality design and prepare templates for the crop formations much like Banksy. They could then execute their crop graffiti through the principle of coincidence, which allows events and occurrences in the hyperphysical domain to overlay the physical level of reality. The principle of energy speed subsets would then enable the hyperphysical artists to operate out of sight to condense super-energy into our world. If our physical third dimension world is overlaid by their hyperphysical fourth dimension world then the fields of Wiltshire would be part of their overall domain. They could overlay a crop with pre-prepared hyperphysical stencils through which superenergy could be condensed to plasticize the crop. Then it could be laid by the energy stream into a masterful image that could appear without them being seen. They wouldn't leave muddy footprints neither would they seek permission from the farmer.

Just like Banksy eyeing a wall the hyperphysical graffiti artists could view a field of rape or corn as a potential canvass. But unlike Banksy they wouldn't use an aerosol can of paint, instead they would deploy a superenergy vortex resonance technology. The equivalent of the Banksy aerosol may have been caught on digital camera when an orb of light was seen dancing over a crop with a pictogram appearing in its wake or it may have been a plasma painting in the plants.

I have described spirits in terms of space plasma. Plasma spheres appearing as orbs have been captured frequently on digital cameras in places such as St Nectans Glen in Cornwall where spirits are thought to gather. The orb pictured forming a crop circle could have been the open end of a superenergy beam delivering a jet of condensing superenergy over the crop. If it was superenergy, in the microwave frequency, condensing through vortex resonance into physical energy over the crop for a fraction of a second that would convert moisture

7 Banksy, Wall and Piece, Century, 2005

in the stems into steam which could plasticize them just long enough so they could be molded and laid by energy streaming over them through an unseen overlay design.

The laying of the crop without damage would appear to be evidence of the application very high temperatures for extremely short intervals of time. Levengood's discoveries demonstrated that this application of thermodynamics was evidently deployed because of the microscopic fused granules of iron found within the soil following the formation of a crop pictogram. The decrease in concentration of these fused particles from the centre of the crop formation outwards is the hall mark of a vortex energy technology.

If only scientists could let go of their attachment to classical materialism and embrace the precedent in quantum theory that everything is energy. If they allowed for the possibility that energy is layered and that we live on one level of energy and people on higher levels can do things in our world because our world is part of their world then maybe evidence from crop formations would be revised and scrutinised by more scientists. Should that occur, they might want to ask where the super-energy artists come from. Do the unseen artists originate from the hyperphysical levels of the Earth or are they associated with UAPs coming in from deep space. Also, what is the purpose of their art. Is it art for art's sake or is it an attempt to communicate with us and maybe broaden our outlook to greater possibilities than are allowed for in the limited, materialistic frame of reference.

CHAPTER 22
UAPS

Since World War II there have been many reports of 'unidentified flying objects,' or *Unidentified Anomalous Phenomena*. Books have been written on the subject and films made but the possibility of extraterrestrial visitors from outer space traveling through our terrestrial skies has been dismissed in mainstream science. In books on the UAP subject, such as *The UFO Phenomenon* by J. Von Buttlar[1] and Timothy Good's *Above Top Secret*[2] the authors claim that reports of UAP's are actively suppressed by the establishment. That makes any serious scientific study of the phenomenon difficult.

Governments have been suppressing evidence of UAPs since 1947 when an alien craft was reported to have crashed near Roswell in New Mexico. Nonetheless, thousands of reports and eye witness accounts, included in innumerable books on the subject, reveal UAPs have been photographed and have appearing on radar both on the ground and in aircraft.

Radar contacts confirm that UAP's are physical objects. At the same time numerous accounts attest to the fact that UAPs vanish quite suddenly and reappear instantaneously. This suggests to me the use of an evaporation and condensation technology to move UAPs in and out of space. If this were so extraterrestrial occupants of UAPs may be deploying super-energy resonance and they may use it to overcome the vast distances of deep space.

Scientists and skeptics dismiss extraterrestrials and UAPs on the grounds that distances in deep space are too vast to be traversed. With the great distances between stars and galaxies in the Universe even at the speed of light, travelling from one planet to another through space would seem to be all but impossible. However, if superenergy resonance is a reality other planetary civilisations might be using this technology to traverse space by moving their spacecraft in

1 Von Buttlar J UFO Phenomenon Sidgwick & Jackson 1979

2 Good T. Above Top Secret, Sidgwick & Jackson, 1987

and out of space rather than through it.

Few people deny the possibility that intelligent beings could be living on planets in other star systems. Amongst the billions of stars and trillions of orbiting planets in our galaxy alone there must be planets with conditions suitable for supporting life. There may be alien civilisations on distant planets that could be millions of years ahead of us in their scientific and technical advancement. If the vortex theory is true it would make sense that somewhere in the Universe intelligent agencies could have developed superenergy resonance technologies for accelerating and decelerating the intrinsic speed of energy in the subatomic matter of their spacecraft to overcome the limitations of travelling through space. Superenergy resonance would enable a spacecraft to be evaporated out of the space of a home planet into the hyperspace, in line with the first cosmic law of motion. In the super-energy resonance beam the craft could then be translocated to Earth. Once in the space of the Earth the beam could be switched off causing the craft to reappear, in line with the second cosmic law of motion.

Moving spacecraft in and out of space-time, rather than through it, opens up the possibility of unrestricted travel for intelligent species in the Universe. Translocating spacecraft from distant planets to the Earth, via hyperspace, might have already happened. Paul Wallis certainly thinks so. In his books *Escaping from Eden*,[3] *The Scars of Eden*[4] and *Echoes of Eden*[5] he presents compelling evidence that extraterrestrials have been visiting the Earth for hundreds of thousands of years and left us behind. Both he and I[6] contend that we are the remnant of a species bred or bioengineered by extraterrestrials as worker slaves. We could be descendants of a cross between their race and terrestrial hominids.

Transposing craft in and out of physical space and time the aliens that made us could have reached Earth space from a distant planet with ease before descended to the surface. The book of Genesis in the Bible speaks of gods descended from space. It also tells that the space gods made us in their own image and after their own likeness before driving us out of their garden because we disobeyed orders.

---

3 Wallis P., Escaping from Eden, Axis Mundi Books, 2020

4 Wallis P., The Scars of Eden, Sixth Books, 2021

5 Wallis P., Echoes of Eden, Paul Wallis Books, 2022

6 Ash D., The Role of Evil in Human Evolution, Kima, 2007

It seems obvious that we are aliens to the Earth because we are so different from any other species on the planet and we have been abandoned because whoever made us has gone while we are still here. They may have gone home not by travelling through deep space but by moving in and out of it. After lifting into hyperspace out of the local space of our planet they could have translocated to the local space of their home planet, dropped back into it and then descended onto its surface, this is possible with superenergy resonance. The novel science I am presenting opens the doors to viable space exploration because it points to a way of overcoming the vast gulfs of space and immense spans of time associated with deep space. Paul Wallis saw a shaman moving about by appearing and disappearing. He called it *pinging*. Super-energy resonance would enable us to ping instantly and anywhere without being constrained by space or time.

Some UAPs could be unmanned reconnaissance vehicles. They may be remotely controlled drones, which have condensed into terrestrial space from mother ships or biosatellites located in deep space. If so they could be tracked by a superenergy resonance beam, and be instantly retrieved if the beam is switched on. The UAP would then vanish by evaporating back into hyperspace. They could then reappear just as suddenly, condensing back into physical space again, if the beam is switched off. This could explain how UAPs can suddenly vanish and then just as quickly reappear again.

Huge UAPs sometimes appear. They could also be condensing into physical space which would account for their sudden appearance and disappearance again due to the deployment of superenergy resonance beams.

Physical craft could incorporate superenergy resonance in their operating systems that would enable them to vanish out of physical space into hyperspace rendering them invisible. Switching off the resonance would immediately cause them to drop back into physical space and reappear again. If I can validate the vortex theory I find it hard to imagine that others have not done so too and already deployed a vortex technology following their own line of scientific enquiry.

CHAPTER 23
EXTRAORDINARY RESEARCH

In the early 1930s, a project of research into invisibility commenced at Chicago University. The team was spearheaded by Nikola Tesla, inventor of AC electric generators and supported by physicists Dr Kurtenhauer and Dr J. Hutchinson, Dean of the University. In 1934 the project moved to Princetown as the Institute of Advanced Study, which included Albert Einstein, a brilliant quantum physicist John von Newmann (who was largely responsible for the development of computers in the USA) and the genius inventor Thomas Townsend Brown.

In 1943 the US Navy gave the go-ahead for the invisibility research team to perform an invisibility experiment on a naval ship moored in the Philadelphia harbour. The apparatus, designed to generate a powerful electro-magnetic vortex, needed to be tested to see if it was possible to make a ship invisible to radar and visible light to help bring about an end to the World War.[1] A destroyer, the USS Eldridge, was chosen for the experiment. Against the advice of the scientists the Navy insisted the crew were left onboard during the experiment.

When the power to the apparatus was switched on the ship vanished from the harbour for fifteen minutes but was reported to have turned up in the harbour of Norfolk Virginia, several hundred miles away, before reappearing in Philadelphia. The experience was catastrophic for the crew. Many of the sailors went out of their minds and five of them were partially reconfigured into the atomic structure of the ship. Hands that had merged into the steel

1 Berlitz C. The Philadelphia Experiment, Souvenir, 1979

where they had been pressed against the bulwarks at the time of the experiment had to be amputated. Needless to say the project was abandoned and classified *Top Secret.*

After the war the invisibility project was renewed, under the direction of Dr Von Newmann, at the Brookhaven National Laboratory until Congress disbanded it in 1967. In *The Montauk Project*,[2] author Preston Nichols claimed that the invisibility experiments were continued from 1971 onwards in secret, as a military project, at the Montauk Air force Base on Long Island, New York, under the continued direction of John Von Newmann. In his book, Nichols claimed that vortex teleportation beams were constructed at Montauk, which enabled people to be moved out of time as well as space. The Philadelphia experiment and the Montauk project suggested that superenergy resonance may have been achieved on the physical plane and used successfully for travelling in and out of space and time. If there is any truth in the Philadelphia experiment and subsequent Montauk experiments they point to the possibility that superenergy resonance may be within reach of terrestrial science and not restricted to hyperphysical operators.

My account for the Philadelphia experiment would be that the vortex generated on board went into vortex resonance with a hyperphysical vortex, which caused the intrinsic speed of energy in every vortex and wave train particle of energy, in the Eldridge and its crew, to equilibrate with the speed of energy in the hyperphysical vortex. That would have caused the ship to evaporate out of physical space and enter hyperspace. For some reason it condensed momentarily into the physical space of Norfolk harbour, evaporated again and then appeared in the grounds of the Montauk air force base on Long Island before evaporating back into hyperspace and condensing back in the harbour in Philadelphia.

In the Philadelphia Experiment, if the USS Eldridge and its crew went beyond the speed of light, then scientists may have stumbled on superenergy resonance. If this was akin to the process deployed by extraterrestrial intelligence to operate UAP craft in and out of the physical space-time of the Earth, then the Philadelphia experiment demonstrated that scientists have already achieved an extraterrestrial level of technology for displacing a craft out of physical space and bringing it back again.

While disastrous for the crew out and about on the ship, according to the reports the technicians in the control room within the ship were unharmed by the experiment. Apparently they benefitted from the protective influence of Helmholtz coils associated with the apparatus they were operating. In the

2 Nichols B. P., The Montauk Project, Sky Books (1992).

subsequent research at Montauk the safety issues were addressed and test subjects were projected regularly in and out of space and time without damage to body or mind.

In *The Montauk Project,* as well as claiming that teleportation beams were constructed at the Montauk Air force Base on Long Island, which enabled bodies to be moved in and out of space and time, Preston Nichols, said that an unintended inter-dimensional link occurred between the Philadelphia experiment in 1943 and a Montauk experiment in 1983, which caused parts of the ship and two crewmembers to be translocated from the USS Eldridge in 1943 to the Montauk base in 1983.

According to Preston Nichols when the USS Eldridge landed on its side at the Montauk Air Force base it left behind at the base some of its superstructure and the two technicians who had been in the control room operating the apparatus inside the ship before it returned to Philadelphia harbour. According to the story when the destroyer landed at Montauk they jumped ship.

The technicians said[3] that when the ship crashed on its side they panicked. Switching off the apparatus they ran from the control room and jumped overboard. They expected to plunge into water but were amazed to land on grass. As they picked themselves up the ship behind them vanished but they found themselves standing alongside some of its superstructure which must have fallen off when it crash-landed. Seeing a group of buildings in the distance they headed off to them. When they arrived they were greeted by surprised personnel including an old man. They were totally bewildered when he introduced himself as the same Dr. John Von Newmann they had seen earlier in the day before the start of the experiment. He was then a man in his prime and now he was old. He was equally shocked. He stared at them in amazement. Here were the two technicians that had gone missing from the Eldridge forty years before.

When they enquired of their whereabouts Von Newmann told them that they were at the Montauk Air Force base on Long Island, then he told them the date. They had leapt off the ship into 1983. It became clear to those working at Montauk that the vortex resonance coils they were working with had taken the ship and two of the men on board, not only out of space but also forward in time.

According to Preston Nichols an experiment conducted at Montauk in 1983 somehow entangled with the Philadelphia experiment in 1943 causing the ship to appear momentarily in Norfolk harbour and then land for a short while at the Montauk base before reappearing in the harbour at Philadelphia. In his

3 Bielek A. & Steiger B. The Philadelphia Experiment, Inner Light

book he said this happened because of a forty year *time-wave*, which caused the experiment on the ship to coincide with an experiment at Montauk. Both experiments involved vortex resonance coils which somehow caused the two experiments to entangle.

My account for what happened was that when the apparatus on the USS Eldridge was switched on and the electromagnetic vortex went into resonance, causing the Eldridge to evaporate into hyperspace, if there were no time wave or Montauk experiment it may have gone invisible but I don't think it would have left the harbour. Had the invisibility research been disbanded, as intended by Congress in 1967, it is likely the ship in 1943 would have simply condensed back into the physical space of Philadelphia harbour. This would have occurred when the power to the vortex resonance coils was switching off. However, the invisibility research didn't stop in 1967. Instead, it continued in secret at Montauk in defiance of Congress. Then on the day in 1983 when the unexpected 40-year coincidence of time waves brought an experiment in Montauk into resonance with the Philadelphia experiment, entanglement happened. I think that an exponential buildup in energy occurred. This can happen with resonance, An uncontrolled build in energy would have caused the ship to be tossed about in hyperspace, as if on a high sea. This and the ship landing in a field could have driven the crew insane. The technicians inside the protective Helmholtz coils, as in the eye of a hurricane, were not affected until the vessel fell onto hard ground. Then frightened, they abandoned ship.

I think it was the chaotic energy that caused the destroyer to condense momentarily into Norfolk harbour and the entanglement caused it to drop into physical space in the grounds of Montauk. The ship must have crashed onto its side when it condensed onto dry land which would obviously terrified the technicians sufficiently to cut the power and run. But the electro-magnetic vortex would not have died immediately. It would have taken a minute or so for the energy in the coils to die away. You may have noticed this effect when the light on a laptop charger takes a few moments to go out after the power is switched off. The delayed decay of power, in the coils in the apparatus on the ship, would have given the two technicians time to make their escape and leap from the vessel to the ground in Montauk. When the power died completely in the coils, the ship would have returned to its original place in Philadelphia harbour in 1943. But the two men who jumped, along with the superstructure that dropped off the ship in the crash, landed in the space-time at Montauk in 1983.

Enabling a ship to fly about, the vortex generating apparatus must have developed an anti-gravity force. This was already familiar to some members of the invisibility team. The enigmatic American inventor Thomas Townsend Brown had been invited to join the Institute of Advanced Studies because had been studying the effect of electric charge on gravity since 1921.[4] He claimed to have discovered that if a disc is charged so the upper side is positive and the lower is negative, when mounted horizontally the disc would thrust upwards toward the positive pole effectively acting against gravity.

In 1929 Brown published a paper on this anti-gravity discovery. He found that a 1% weight loss could be generated by a 100Kvolt electric field. He went on to invent a revolutionary anti-gravity motor with no moving parts. His anti-gravity disc was segmented so that each segment could be selectively charged. Moving the charge around the rim of the disc from one segment to another allowed the anti-gravity force to be directed. The implication of this discovery for aircraft was phenomenal.

In the mid 1970's, the English inventor of the monorail and emeritus professor of heavy engineering at Imperial College London, Eric Laithwaite, discovered the anti-gravity effect of gyroscopic spin. He was able to wheel gyroscopes around his head with only one hand when they were spinning but when stationary they were too heavy to lift. Eric Laithwaite is reported to have said, *"I believe matter itself is just spin."*

In 1984 a Scottish inventor, Sandy Kidd, built a device with gyroscopes spinning at each end of a cross bar. When the gyros were spinning the apparatus levitated a few inches off the bench. I saw this demonstrated on television.

John R. Searl was an electrical engineer, employed by the Midland Electricity Board who investigated an anti-gravity force generated by spinning electric fields. He set a segmented rotor disc spinning through electro-magnets at its periphery. The electromagnets, energized from the rotor, were intended to boost the electromotive force. The generator, about three-foot in diameter, was first tested in the open by Searl and a friend in 1952. To begin with it produced the expected electric power but an unexpectedly high voltage. This quickly exceeded a million volts producing a crackling sound and the smell of ozone. In Searl's own words: *"Once the machine has passed a certain threshold potential the energy output exceeded the input. From then on the energy output seemed to be virtually limitless."* [5] Then as the generator continued to increase in potential it lifted off the ground and broke free of its mountings and the engine. It floated in the air all the time spinning faster as the air around it glowed pink with ionisation and it then accelerated off into space and was

4 The Biography of Thomas Townsend Brown & Project Winterhaven The Townsend Brown Foundation

5 Wynniatt C.B., Energy Unlimited, Issue 20, 1986

never seen again. In subsequent experiments Searl mounted his generators, which he built up to thirty foot in diameter, more firmly in the ground. But they still tore themselves free of the Earth taking the foundations with them. The hemispherical crater left in the ground suggested the anti-gravity force was operating over a sphere with the generator at its centre.

An Austrian inventor Victor Schauberger, famous for his construction of logging flumes, discovered the anti-gravity potential of vortex motion by chance. Victor Schauberger was a young ranger in the wilderness forest of Bernerau, in Austria, when he made his first observations of the power in vortex motion. In his own words: *"It was spawning time one early spring moonlight night. I was sitting by a waterfall waiting to catch a fish poacher. What then occurred took place so quickly that I was hardly able to comprehend. In the moonlight falling directly onto the crystal clear water, every movement of the fish, gathered in large numbers could be observed. Suddenly the trout dispersed due to the appearance of a particularly large fish, which swam up from below to confront the waterfall. It seemed as if it wished to disturb the other trout and danced in great twisting movements in the undulating water as it swam quickly to and fro. Then as suddenly the large trout disappeared in the jet of the waterfall which glistened like falling metal. I saw it fleetingly under a conical-shaped stream of water, dancing in a wild spinning movement the reason for which was not at first clear to me. It then came out of this spinning movement and floated motionlessly upwards. On reaching the lower curve of the waterfall it tumbled over and with a strong push reached behind the upper curve of the waterfall. Deep in thought I filled my pipe and as I wended my way homewards, smoked it to the end. I often subsequently saw the same sequence of play of a trout jumping a high waterfall."* [6]

Schauberger observed that the vortex motion of water, a little above freezing, helped to lift trout up waterfalls. He was also intrigued by the way trout in the mountain streams would remain motionless as if suspended in the fast-flowing water then dart upstream. Schauberger was convinced that the turbulent

---

6 Alexandersson Olaf, Living Water: Viktor Schauberger and the Secrets of Natural Energy, Gateway Books, (1990).

motion of water, at its greatest density, generated a force in the opposite direction to the flow of the stream. He believed that trout could seek out the upstream flow of energy and use it to remain motionlessly in the fast flow of water or to propel them upstream and over waterfalls.

Victor Schauberger believed that a trout could take advantage of a force generated by the spiral motion of water passing from its gills over the surface of its body. Convinced the vortex was a source of energy he decided to test his idea by building a vortex turbine based on the same principle of twisting, reeling and spinning that he had observed in the fast-flowing waters of freezing mountain streams. His most successful designs were based on the corkscrew shaped spirals expelled from the gills of trout so he called his apparatus the trout turbine.

Schauberger was remarkable for his ability to construct heavy apparatus without any training, engineering facilities or funding. Nonetheless he was successful and reported that success depended on the temperature of the water, the shape of the turbine and materials of which it was constructed.

In the early 1930's Schauberger fabricated conical pipes of special materials, which contained a corkscrew turbine. Operated by an electric motor the spiral turbines screwed water into a vortex flow and directed the water onto a conventional water turbine coupled to a generator. He claimed that as the water was screwed faster it suddenly began to produce large amounts of energy. The turbine began to generate more electricity than the input motor was using. Then suddenly the system went out of control. The apparatus tore itself away from its holdings and smashed against the ceiling. When Schauberger experimented with air turbines he found the same thing happened. Regardless of the medium it was in, vortex motion seemed to cross a threshold then generate energy, apparently out of nowhere, which also produced an anti-gravity force.

Just before the outbreak of the World War II apparently Adolph Hitler took an interest in Schauberger's work. [5] He ordered a Vienna firm called Kertl to construct and test Schauberger's vortex turbines with a view to using them in aircraft engines. An engineer called Aloys Kokaly was employed in the manufacture of parts. On one occasion when he delivered parts to the Kertl factory he was told, *"This must be prepared for Mr. Schauberger on orders from higher authority, but when it's finished, it's going out onto the street, because on an earlier test on one of these strange contraptions, it went right through the roof of the factory."* [6]

Another innovator who may have stumbled on vortex resonance while he was experimenting with spin was an American inventor, Joseph Newman[7]. Newman found that energy could be produced by setting electromagnetic fields spinning. His machine consisted of a number of rotating magnets wound with copper wire to form a reciprocating magnetic armature. According to Newman as the armature was set spinning, an electromagnetic force was induced which set up a spiral pattern of motion around the current carrying copper wire. The apparatus then began to produce energy out of nowhere.

In *The Guardian* newspaper it was reported that Dr Roger Hastings, chief physicist for the *Sperry-Univac* Corporation, tested Newman's apparatus.[8] He found the production efficiency of the machine to be far greater than 100%. On September 20th 1985 Hastings issued an affidavit to the effect that "...On September 19th 1985 the motor was operated at 1,000 and 2,000 volts battery input, with output powers of 50 and 100 watts respectively. Input power in these tests were, 7 and 14 watts yielding efficiencies of 700% and 1,400% respectively..."

These experiments appear to have discovered a source of anti-gravity and free energy and it could be a consequence of a resonance associated with vortex motion but they didn't prove that. The theory of vortex resonance is just a rationale I developed to help me understand this extraordinary research. Nonetheless, I do believe that vortex resonance could account for the over unity energy output of this amazing research. My postulate is that a vortex in physical matter could resonate with a similar form of vortex in hyperphysical matter. I think it is the similitude of shape that would allow superenergy resonance to occur between vortices in the physical and hyperphysical dimensions. If a vortex is set up in a medium on the physical plane, albeit in water, air or in an electromagnetic field, it could match up with a vortex of similar form in the hyperphysical realm. With no space-time separation between the matching vortices in the physical and hyperphysical dimensions resonance could occur between them.

The principle of energy speed subsets suggests that the direction of energy flow would be from the greater to the lesser. The self-evident laws that 'water flows downhill' and 'electricity flows down potential gradients' supports my prediction that in vortex resonance between physical and hyper-physical systems, the energy would flow from the hyperphysical system into the physical. I believe if a vortex is set in motion and its speed is steadily increased, at a certain threshold speed resonance should occur. Once resonance starts the

7 Newman Joseph Westley, The Energy Machine of Joseph Newman, (1986) Lucedale, Mississippi USA.

8 The Guardian, March 21, 1986.

energy in the vortex and the anti-gravity force appear to increase exponentially. That would account for the apparatus losing control, ripping free of its holdings and then literally going through the roof.

The possibility of unlimited energy output is a feature of resonance. This is clear from radio and television where there is no limit to the number of televisions or radios that can tune in and resonate with a single broadcast signal. With no space-time separation between physical energy and super-energy systems, theoretically a physical vortex could resonate with every single matching vortex of superenergy in existence. By doing so the physical vortex could, through vortex resonance, draw virtually unlimited amounts of energy into our world from higher dimensions. This could be very dangerous and counter spinning vortices would need to be set up to damp the system and maintain control - much as an air throttle is needed to damp down and control an internal combustion engine.

Vortex resonance has the potential to provide unlimited power. At the same time, it could be very destructive if mishandled or misused. Maybe it is fortunate superenergy technologies have not been entertained in mainstream science and engineering. Everything that happens in the world is a reflection of the consciousness of humanity. Because we have not yet matured to the level of responsibility required for the quantum leap in technology that vortex resonance will bring, resistance to developing the vortex technology that frustrates progress may be a blessing in disguise.

Conspiracy theories abound in regard to free energy and anti-gravity research but we should not be over concerned with conspiracies. We should be more concerned with consciousness. It is the rift between science and spirituality that is frustrating the consciousness shift necessary before we are ready to develop vortex and superenergy resonance without further threat to our survival. In my opinion we should turn our attention from technology to personal development, spiritual evolution and to our ascension which is the ultimate goal of superenergy resonance. If we achieve that then freedom of the Universe will be ours to enjoy.

CHAPTER 24
ASCENSION

In 1991, as I embarked on my first lecture tour in Australia, I was introduced to a set of five cassette tape recordings predicting a spiritual process called *ascension.* A Californian, by name of Eric Klein, had made a set of compelling voice recordings, which he *channeled* from the fifth dimension. Subsequently published in his book, *The Crystal Stair,*[1] the channelings detailed what the predicted ascension was, how it would occur and how humanity could best prepare for it. I trusted them because Eric channeled Jesus Christ and my heart told me they were true. I am multifaith and no longer endorse the trinity but I do believe Jesus was one of the most authentic men in history and I have implicit faith in his role as a planetary guardian of humanity. Because of the apparition experience I had in Medjugorje, when I listened to the start of the recording of his first channeling I recognised immediately who it was speaking through Eric Klein.

I accepted the ascension information, first introduced to Eric Klein by Canadian, Ariana Sheran in 1990, because I had predicted the process that could account for it in the book I wrote with Peter Hewitt.[2] Consequently, I devoted the next four years of lecturing around the world to propagating

1 Klein E. The Crystal Stair, Oughten House, 1992

2 Ash D. & Hewitt P. The Vortex: Key to Future Science, Gateway Books, 1990

a fusion of the vortex physics and transformative spirituality based on the ascension information. The message from the superphysical, fifth dimensional level of intelligence speaking through the physical conduits in our third dimension was that humanity is now living in a singularly unique period of planetary history. We are going to go through an immense spiritual transition prior to the planet undergoing ascension into the superphysical fifth dimension. In the Eric Klein channelings it was explained that there would be three waves of ascension giving everyone an equal opportunity to ascend from the third dimension to the fifth dimension. Those who ascend in the first wave would be offered the opportunity to return to the Earth to help people make it through the subsequent two waves. Those who return will have the power to heal and work miracles. The second and third waves will precede a cleansing of the Earth prior to its ascension. The third wave, presenting the last opportunity for people to ascend, will include an evacuation of children and innocents. An assurance was given for those who choose to ascend that children in their care will be evacuated with them at their time of ascension.

In the Eric Klein channelings, Jesus used his Vedic name *Sananda* as well as his Christian name *Jesus*. He described the ascension process as follows: *"So what is the [ascension] process? You have grown accustomed to reincarnational experience – that is, having your soul incarnate in a body and going through a lifetime of a certain number of years, experiencing death, and leaving your body to return again to the earth. You have gone to some fourth-dimensional areas. There are heavens and hells galore, with many experiences. Yet always you return to the body, for the body is the platform from which your launch will take place into the fifth dimension. This is the intention, let me say, of having a physical body, at this time especially. I would say that your ascension will be all but identical to my own. You will not leave your bodies behind and go to a higher state of consciousness. Your bodies will be transformed also. The molecules and atoms, your subatomic particles, all that you are, will be transformed and accelerated into the fifth dimension. So you do not have to die. Well, that's some good news. Despite the fact that human beings have grown so accustomed to dying that it has become a common awareness or common belief, I am telling you now: you do not have to die. And you will not."* [1]

Many seers predict we are in a period prior to the Earth going through massive changes. This prediction was given in the Marian apparitions at Garabandal [3] in Spain in the 1960's. It was said there that there would come a time when the Earth will be chastised by fire, which would be preceded by a warning, which everybody on Earth would witness. The warning would then be followed by a

3 Sánchez-Ventura y Pascual F., The Apparitions of Garabandal, San Miguel, 1966

miracle to help people believe.

In the Eric Klein voice recordings and in his follow on book[1] it was predicted that human beings will be offered the opportunity for evacuation and ascension, prior to planetary purification, through the experience of a *door of light.* When the door of light appears it will be without warning. Each of us will have the same opportunity to enter into it. Whether awake or asleep we will all be fully conscious of the door of light. It will linger for a minute or two. If we hesitate or choose not to enter it will fade away. In the channelings it was made clear that *the elect* are not an elected few but are the few who elect to enter the door of light when it is offered to them. If the ascension predictions are true, entry into the fifth dimension will be a process of self-selection. We each have to decide for ourselves. According to David Wilcock to be cleared for ascension we need to be just over 50% in service to others over being in service to ourselves.[4]

On my lecture tour I taught that civilisation with all its distractions and temptations is acting as an *ascension filter.* Never in history have there been so many things for people to become attached to on the physical plane. When the unique opportunity for ascension is offered to us we will have to choose immediately what is more important to us, the spiritual or the material. When the invitation to ascension opens before us as a door of light we will be called on to let go of everything we are attached to in the physical world if we want to step through the door to eternity into the spiritual world. It will be a test of priority between the spiritual and the material.

The vast majority of people who ever lived have died. Only a tiny minority have ascended. At death we have to leave everything behind. With death there is no choice. In order to ascend have to choose. Will we be willing to step into the unknown and leave behind everyone and everything we know.

Ascension is not an easy choice. It requires faith and detachment, trust, spontaneity, fearlessness and courage. In order to ascend a person needs to prioritise selfless service over selfish attachment and be more in their heart than in their head as the heart is attracted to light and freedom whereas the head is more attracted to sensibility, security and to what is known and what is familiar. Death is familiar. Those who do not elect to ascend will die eventually because death is the normal process.

In the Eric Klein channelings it was said that at the time of the third wave souls will ascend from the fourth dimension where people go after death. No one will miss out on ascension if that is their desire and if they are ready to

4 Wilcock D., The Ascension Mysteries, Souvenir Press, 2017

ascend.

Most people are only concerned with their physical body and life on the physical plane but near-death experiences give us a glimpse of a world beyond the physical, which seems certain for us all. The widely held materialist assumption that nothing exists apart from physical reality and fear of the unknown act as blocks to ascension.

Ascension is unprecedented. It is not the normal way to depart the Earth. It is not easy to accept it as possible but in my theory of superenergy resonance I can explain how ascension of the physical body could occur, though I cannot say for sure when or even if it will occur. I can only suggest how ascension might occur for you if it does happen.

The door of light, acting as an invitation to ascension, could be the end of a superenergy resonance beam. If so as each one of us steps into this doorway of light, or desires to enter it in our minds, we would be allowing ourselves to enter the beam of superenergy. Should we do so every single particle of energy in our body would resonate with the superenergy in the beam. The intrinsic speed of energy in our physical body would accelerate to match the speed of superenergy in the beam. We would then experience a quickening and find ourselves travelling up a tunnel of light into a higher dimensional state.

According to the cosmic laws of motion, if we are to ascend into the fifth dimension with our bodies we would need a fifth dimensional planet for them to live on so planetary and bodily ascension must go hand in hand. I began to understand the process of planetary ascension, during my world lecture tour in 1992/93, when it became clear to me the solar system could be in a region of space where the speed of light is depressed, which I called a *zone of density.* As a boy at Westminster Cathedral Choir School I was taught we went through a *fall in grace* at the time of Adam and Eve. On my tour I learnt this was about eleven thousand years ago. In 1994 I read of catastrophic Earth changes between eleven and twelve thousand years ago.[5] This may have been linked to the solar system entering a region of space contaminated by a cosmic cataclysm that set up a zone of density in which the speed of energy in the region may have become depressed to the speed we now measure as the speed of light. I related grace to the intrinsic speed of energy in matter and light so for me religious predictions of a time of grace coming when the fall would be reversed were synonymous with the predictions of ascension.

In my lectures I explained that the speed of energy in the Universe is mostly

---

5 Allen D. & Delair J. When the Earth Nearly Died: Compelling evidence of a Catastrophic World Change 9500B.C. Gateway Books, 1994

faster than the speed of light but when photons of light from distant stars enter a zone of density the speed could be depressed, much as the speed of light drops as it enters glass or water. Just as the speed of light reverts to normal as it leaves the glass or water so on leaving a zone of density the starlight would revert to its normal higher speed. I then explained that the planet we are on is about to leave a zone of density when the intrinsic speed of energy in every subatomic vortex and photon of light will revert naturally to the higher speed of energy as it was before the solar system entered the zone. That, I predicted, is how planetary ascension would occur.

While the zone of density was conjecture it made sense of the concept of *quarantine* in the ascension movement. In metaphysics there is a teaching that the fall of grace led to our quarantine because of our selfish tendencies, which manifest as fear, greed, conflict and war. We are in quarantine because these tendencies pose a threat to the Universe. In my talks I said that evacuation by ascension was necessary before we leave the zone of density so the Earth can be cleansed of pollution and parasites, which include people inclined to selfishness, hatred, greed and war; people who take more than they give. Before the planetary purification there would be an evacuation of plants, animals and humans inclined to peace, love and selflessness. They would be lifted out of danger into a holding station in a higher dimension so that along with benign life they can be returned to the planet after it is cleansed, prior to its leaving the zone of density.

CHAPTER 25
THE PETER DEUNOV PROPHECY

In 2017 Matthew Newsome introduced me to the prophecy of the Bulgarian Master, Peter Deunov which corroborated many of the things I had been talking about in regard to ascension.[1]

Albert Einstein said of Peter Deunov, *"The whole world bows down before me; I bow down before the master Peter Deunov from Bulgaria"*. Pope John XXIII described Peter Deunov as, *"The greatest philosopher on Earth."* Days before he died Peter Deunov (1864-1944) came out with an astounding prophecy which ratified much of what I have said in regard to human and planetary ascension. Here is what the Master had to say:

During the passage of time, the consciousness of man traversed a very long period of obscurity. This phase, which the Hindus call 'Kali Yuga', is on the verge of ending. We find ourselves today at the frontier between two epochs: that of Kali Yuga and that of the New Era that we are entering.

A gradual improvement is already occurring in the thoughts, sentiments and acts of humans, but everybody will soon be subjugated to divine Fire, that will purify and prepare them in regards to the New Era. Thus man will raise himself to a superior degree of consciousness, indispensable to his entrance to the New Life. That is what one understands by 'Ascension'.

---

1 . Kraleva M. The Master Peter Deunov: His Life and Teaching, Kibea 2001

Some decades will pass before this Fire will come, that will transform the world by bringing it a new moral. This immense wave comes from cosmic space and will inundate the entire earth. All those that attempt to oppose it will be carried off and transferred elsewhere.

Although the inhabitants of this planet do not all find themselves at the same degree of evolution, the new wave will be felt by each one of us. And this transformation will not only touch the Earth, but the ensemble of the entire Cosmos.

The best and only thing that man can do now is to turn towards God and improve himself consciously, to elevate his vibratory level, so as to find himself in harmony with the powerful wave that will soon submerge him.

The Fire of which I speak, that accompanies the new conditions offered to our planet, will rejuvenate, purify, reconstruct everything: the matter will be refined, your hearts will be liberated from anguish, troubles, incertitude, and they will become luminous; everything will be improved, elevated; the thoughts, sentiments and negative acts will be consumed and destroyed.

Your present life is a slavery, a heavy prison. Understand your situation and liberate yourself from it. I tell you this: exit from your prison! It is really sorry to see so much misleading, so much suffering, so much incapacity to understand where one's true happiness lies.

Everything that is around you will soon collapse and disappear. Nothing will be left of this civilization nor its perversity; the entire earth will be shaken and no trace will be left of this erroneous culture that maintains men under the yoke of ignorance. Earthquakes are not only mechanical phenomena, their goal is also to awaken the intellect and the heart of humans, so that they liberate themselves from their errors and their follies and that they understand that they are not the only ones in the Universe.

Our solar system is now traversing a region of the Cosmos where a constellation that was destroyed left its mark, its dust. This crossing of a contaminated space is a source of poisoning, not only for the inhabitants of the earth, but for all the inhabitants of the other planets of our galaxy. Only the suns are not affected by the influence of this hostile environment. This region is called 'the thirteenth zone'; one also calls it 'the zone of contradictions'. Our planet was enclosed in this region for thousands of years, but finally we are approaching the exit of this space of darkness and we are on the point of attaining a more spiritual region, where more evolved beings live.

The earth is now following an ascending movement and everyone should force themselves to harmonize with the currents of the ascension. Those who refuse to subjugate themselves to this orientation will lose the advantage of good conditions that are offered in the future to elevate themselves. They will remain behind in evolution and must wait tens of millions of years for the coming of a new ascending wave.

The Earth, the Solar system, the Universe, all are being put in a new direction under the impulsion of Love. Most of you still consider Love as a derisory force, but in reality, it is the greatest of all forces! Money and power continue to be venerated as if the course of your life depended upon it. In the future, all will be subjugated to Love and all will serve it. But it is through suffering and difficulties that the consciousness of man will be awakened.

The terrible predictions of the prophet Daniel written in the bible relate to the epoch that is opening. There will be floods, hurricanes, gigantic fires and earthquakes that will sweep away everything. Blood will flow in abundance. There will be revolutions; terrible explosions will resound in numerous regions of the earth. There where there is earth, water will come, and there where there is water, earth will come. God is Love; yet we are dealing here with a chastisement, a reply by Nature against the crimes perpetrated by man since the night of time against his Mother; the Earth.

After these sufferings, those that will be saved will know the Golden Age, harmony and unlimited beauty. Thus keep your peace and your faith when the time comes for suffering because it is written that not a hair will fall from the head of the just. Don't be discouraged simply follow your work of personal perfection.

You have no idea of the grandiose future that awaits you. A New Earth will soon see day. In a few decades the work will be less exacting, and each one will have the time to consecrate spiritual, intellectual and artistic activities. The question of rapport between man and woman will be finally resolved in harmony; each one having the possibility of following their aspirations. The relations of couples will be founded on reciprocal respect and esteem. Humans will voyage through the different planes of space and breakthrough intergalactic space. They will study their functioning and will rapidly be able to know the Divine World, to fusion with the Head of the Universe.

The New Era is that of the sixth race. Your predestination is to prepare yourself for it, to welcome it and to live it. The sixth race will build itself around the idea of Fraternity. There will be no more conflicts of personal interests; the single aspiration of each one will be to conform himself to the Law of Love. The sixth race will be that of Love. A new continent will be formed for it. It will emerge from the Pacific, so that the Most High can finally establish His place on this planet.

The founders of this new civilization, I call them 'Brothers of Humanity' or also 'Children of Love'. They will be unshakeable for the good and they will represent a new type of men. Men will form a family, as a large body, and each people will represent an organ in this body. In the new race, Love will manifest in such a perfect manner, that today's man can only have a very vague idea.

The earth will remain a terrain favourable to struggle, but the forces of darkness will retreat and the earth will be liberated from them. Humans seeing that there is no other path will engage themselves to the path of the New Life, that of salvation. In their senseless pride, some will, to the end hope to continue on earth a life that the Divine Order condemns, but each one will finish by understanding that the direction of the world doesn't belong to them.

A new culture will see the light of day, it will rest on three principal foundations: the elevation of woman, the elevation of the meek and humble, and the protection of the rights of man.

The light, the good, and justice will triumph; it is just a question of time. The religions should be purified. Each contains a particle of the Teaching of the Masters of Light, but obscured by the incessant supply of human deviation. All the believers will have to unite and to put themselves in agreement with one principal, that of placing Love as the base of all belief, whatever it may be. Love and Fraternity that is the common base! The earth will soon be swept by extraordinary rapid waves of Cosmic Electricity. A few decades from now beings who are bad and lead others astray will not be able to support their intensity. They will thus be absorbed by Cosmic Fire that will consume the bad that they possess. Then they will repent because it is written that "each flesh shall glorify God".

Our mother, the earth, will get rid of men that don't accept the New Life. She will reject them like damaged fruit. They will soon not be able to reincarnate on this planet; criminals included. Only those that possess Love in them will remain.

There is not any place on earth that is not dirtied with human or animal blood; she must therefore submit to a purification. And it is for this that certain continents will be immersed while others will surface. Men do not suspect to what dangers they are menaced by. They continue to pursue futile objectives and to seek pleasure. On the contrary those of the sixth race will be conscious of the dignity of their role and respectful of each one's liberty. They will nourish themselves exclusively from products of the vegetal realm. Their ideas will have the power to circulate freely as the air and light of our days.

The words "If you are not born again" apply to the sixth race. Read Chapter 60 of Isaiah it relates to the coming of the sixth race, the Race of Love.

After the Tribulations, men will cease to sin and will find again the path of virtue. The climate of our planet will be moderated everywhere and brutal variations will no longer exist. The air will once again become pure, the same for water. The parasites will disappear. Men will remember their previous incarnations and they will feel the pleasure of noticing that they are finally liberated from their previous condition.

In the same manner that one gets rid of the parasites and dead leaves on the vine, so act the evolved Beings to prepare men to serve the God of Love. They give to them good conditions to grow and to develop themselves, and to those that want to listen to them, they say: "Do not be afraid! Still a little more time and everything will be all right; you are on the good path. May he that wants to enter in the New Culture study, consciously work and prepare."

Thanks to the idea of Fraternity, the earth will become a blessed place, and that will not wait. But before, great sufferings will be sent to awaken the consciousness. Sins accumulated for thousands of years must be redeemed. The ardent wave emanating from On High will contribute in liquidating the karma of peoples. The liberation can no longer be postponed. Humanity must prepare itself for great trials that are inescapable and are coming to bring an end to egoism.

Under the earth, something extraordinary is preparing itself. A revolution that is grandiose and completely inconceivable will manifest itself soon in nature. God has decided to redress the earth, and He will do it! It is the end of an epoch; a new order will substitute the old, an order in which Love will reign on earth."

Deep within the human heart
Lies the spring of life, the eternal spark,
The door to heaven stands open wide,
The door is in the heart so step inside,
Step within the breath you breathe,
Breathe in deep now and feel the peace,
Let the breath lead you into your heart,
Let it connect you to your eternal spark.

The plan of ages slowly unfolds,
Within its great purpose man evolves,
Ages of suffering grim to bear,
But how else would we learn to love, care and share,
So let not your problems hold you in their sway,
Nor let your troubles sweep you away,
These are just the trials that make us strong,
They'll keep on coming while to the Earth we belong.

So breathe your way through every trial and task
Knowing that none will ever last,
And breathe into every good fortune and joy,
Knowing also that they will pass,
Our lessons they come in light and dark,
Let neither trick you from your eternal spark,
Good and evil are but the masks,
That life wears in the play of human hearts.

BOOK II

THE PLAN OF AGES

"...life isn't easy, but maybe it isn't meant to be easy... I realise that we don't always have to seek out the easiest path, or take the one that's presented to us; sometimes its's the hardest one that holds the greatest riches."

- Raynor Winn
Landlines p.296
Penguin 2022

INTRODUCTION

Academics and materialists like Yuval Noah Harari would have us believe that *"In itself the Universe is only a meaningless hodgepodge of atoms,"*[1] and there is no purpose to human life apart from individual and species survival. But materialism is just a story. You can embrace the story of scientific materialism or ditch it as you choose. It is entirely up to you. You hold your destiny in your own hands and you are free to believe whatever feels right to you.

Feel into my story. Does it feel true to you? Does it resonate with your innermost being. Do you trust yourself because while you are in the Universe, the Universe is in you and from that wellspring of innate knowledge you will know what is right and what is wrong. You can sharpen your power of discernment by trusting your intuitive knowing. Any trickster who speaks otherwise is attempting to steal your power so you believe their story instead of your own gnosis. In my story you are an eternal being, a master of your own destiny. You can determine your own future but remember what feels right to you may not resonate with others. Rigid beliefs in science or religion can be dangerous and demonising others because of their beliefs, even more so. Toss my ideas in the air. Feel into them allowing only those that feel right to land in your heart. Have fun in the process and be in your joy as you journey into new possibilities.

On my journey of discovery I chose to move seamlessly between science and religion, folklore, fable and physics. Sifting fact from fiction against the benchmark of superenergy and the quantum vortex I found the mysteries of spirituality, the supernatural and miracles unveiled and elements of truth were revealed in traditional mythologies. Then a coherent pattern began to emerge that pointed to an incredible future for us all in which we have an opportunity to be included in the creation and destiny of the Universe. Far from being pointless I discovered there is an immense purpose to human life which is unfolding before our eyes. I believe we are participants in a plan of ages that is slowly unfolding toward an end point which, for you and I, is happening right now.

I have a deep love for humanity and as an eternal optimist I have unshakable faith that our individual future and the future of all humankind will be truly amazing. All my life I have felt a great change is coming, a massive spiritual transition into a new age of wisdom and joy for us all, where there will be

---

1 Harari Y.N. 21 Lessons for the 21st. Century, Jonathan Cape, 2018

less inhumanity and injustice and more love in the world.In order for this to occur the Earth needs to undergo a refit in the form of unavoidable changes and necessary regeneration. The planet and humanity need to be rewilded. We and the planet need to be released from the burden of industrial, political and corporate dominion so we can find our own future in a world where there is true freedom. We are offered ascension and evacuation while the Earth is purified by natural forces, then we will have the opportunity to repopulate our beloved planet after she has been restored to her natural, pristine state.

The teachings about this transition make it clear an optimistic future is available to everyone. We do not have to be special, spiritually advanced or have any particular religious belief. All we need is faith in the Great Spirit and to live and act out of unconditional love. To be less and have less means we will have less to let go of - the meek shall inherit the Earth. If our focus is on being rather than achieving; if we are happy to be nobody with nothing but a good heart, then like the caterpillar that becomes mush before metamorphosing into a butterfly, we will more easily enter the void we must pass through in order to undergo our transition into the light. Can we be totally committed to our lives as they are now and then if called, in an instant of joy, let go in spontaneous abandonment to divine love and without 'thinking about it', relinquish hold on all things of this world and step into the light.

How we choose to live, day by day, determines not only our destiny in this world but also in the world to come. If we live out of love rather than fear, surrendering criticism and judgment of others and also of ourselves, if we manage to overcome our attachment to people, places and things then we will be better primed to let go and enter the opening of light to be a beneficiary of this planet's ascension process. This is not a call to give up our possessions, our jobs or our homes. It is a call to release our attachment to them. We should fulfill our daily commitments and live life to the full, but as ***Ry Cooder*** sang, "If you'd like to get to heaven and see eternity unfold, You must, you must unload." [2]

---

2 Cooder R., The Prodigal Son, Track. 7, You Must Unload, Perro Verde Recordings LLC, 2018

CHAPTER 1
THE WOMB OF ANGELS

Many of us question the meaning and purpose of human life but an answer to our questions came, in the 18th century, through a remarkable Swede. Emanuel Swedenborg was a well-educated, widely traveled Swede who went through a profound spiritual awakening in 1745. It occurred during a health crisis from which he nearly died. He said he had been to another dimension where a heavenly guide gave him revelations for humanity. After his crisis he continued to have spiritual visions and dreams. Swedenborg had already established himself internationally as a scientist, inventor, mathematician and philosopher so he was taken seriously by a great number of people.

The teachings of Swedenborg[1] had a profound influence on Blake, Goethe, Emerson, Dostoevsky, the French Symbolists, Kant and Jung. The essence of his message was that human beings exist simultaneously in physical and spiritual worlds. After death, memories of the physical world fade whereas those of the spiritual world survive. He said Heaven is much like the earth except that people live in spiritual bodies rather than physical bodies and they are able to enjoy pleasures, including sex, but life there is comparatively mundane. This fits with my premise that we live simultaneously in a hyperphysical body in parallel to our physical body, which separates when the physical body dies and goes on living in the greater hyperphysical world.

Swedenborg was also taught that Hell exists but there are no devils or Satan there. The core of Swedenborg's revelation is that God, Heaven and Hell exist within us. He was told that we should ditch doctrines like the Trinity and the idea that Christ died on the cross to atone for the sins of humankind. He was told that on the spiritual plane salvation comes through individual goodness and personal striving to live a spiritual life.

---

1 Swedenborg E., A Swedenborg Sampler: Selections from Heaven and Hell, Divine Love and Wisdom, Divine Providence, True Christianity, and Secrets of Heaven, Swedenborg Foundation Publishers, 2011

Emmanuel Swedenborg devoted the last twenty seven years of his life to teaching and writing. His most famous books were *Arcana Ceolestias*,[2] and *Heaven and Hell*.[3] One of the most significant points in Swedenborg's heterodoxy is that: *The human race is the basis on which heaven is founded and angels arise from the souls of humans.*

In *The World of Angels*[4] is written: *Contrary to orthodox theology, Swedenborg believed that angels are not created in heaven by God but arise from the souls of deceased human beings. In the Swedenborg celestial hierarchy, a soul's place in angelic society is determined by his or her beliefs and sensibilities as an earthly mortal.*

This teaching of Emanuel Swedenborg has profound implications. If Swedenborg is right then how we live right now determines our eternal destiny. According to his profound insights the Earth is, in effect, a *Womb of Angels*. Just think about it. The planet we are living on could be the place where angels are germinated, gestated, trained and upgraded. We may be angels in the making.

The idea that the Earth is a womb of angels is powerful. It casts a positive light on death as a time of birth into a new life experience. This has been confirmed by recent research into near death experiences. When we were born we passed through the tunnel of the birth canal into the light of the physical world. Evidence from near death experiences confirm when we leave this world we pass through a tunnel into the light of the next world.

I don't believe we complete the process of becoming an angel in a single lifetime. I believe we graduate as an angel on our ascension. Swedenborg confirmed this. The idea we have only one life or that we are born as an angel after we die could be a simplification of a much wider picture. Research into reincarnation suggests that when we die we are born into another world from whence, for those who reincarnate, there is a return to another life here in a new physical body. I denoted the place we go to after physical death as the hyperphysical plane. Others call it the fourth dimension. Whatever we call it, it would seem the world we enter after death is a resting station between the many incarnations we need to prepare us for our eternal destiny in the angelic hierarchy.

I teach that the hyperphysical plane is integral to the Earth. It is the Earth Soul or *Anima Mundi*. I believe it is where we go to after physical death to

2 Swedenborg E. Arcana Coelestia, Forgotten Books, 2008

3 Swedenborg E. Heaven and Hell, Swedenborg Foundation Publishers, 2010

4 Penwyche G. The World of Angels Bounty Books 2009

recuperate and integrate our terrestrial experience. Nonetheless, if we live an exemplary selfless life and make great sacrifices for the betterment of humanity and the world we may, after death, ascend beyond the hyperphysical, into the superphysical planes of the Universe - the fifth dimension and beyond - where the angelic and divine intelligences dwell. We may graduate as an angel or *ascended master* after any lifetime if we are ready. Ascended humans are known as *Ascended Masters,* not because they have become masters of others but because they have become masters of themselves.

Should someone be ready to ascend after an exceptional lifetime, it would be in the wake of many incarnations on Earth learning and training as an angel to be. During the hyperphysical interim between incarnations the fruits of physical life are pressed through the hyperphysical wine-press. Each soul finds itself attracted by frequency resonance, according to its nature, desires and actions in the physical world, to a heavenly or hellish situation in the hyperphysical. Both outcomes serve in the progress of angelic advancement. These polarity situations are self-imposed. As Immanuel Swedenborg said, *"Heaven and hell are from the human race."*

The heavens and hells in the hyperphysical dimensions present opportunities for souls to review actions chosen during physical incarnation. Crucially they provide an opportunity to review the consequences of these actions, as they reverberated through the lives of other people.

The experience of joy or sorrow felt by a soul in the 'after-life' is determined by how much that soul affirmed or negated itself in this life. Joy or sorrow in the hyperphysical heavens and hells are hi-fidelity states for the soul because the hyperphysical body is very heightened in sensation when it is free of the damping influence of the physical body. Both polarities provide necessary feedback for the soul individualisation of spirit in its evolution into mastery. The soul individualisation of spirit is often referred to as the 'Higher Self', the 'Divine portion', or the 'I AM' presence.

As the Earth progresses toward planetary and mass human ascension learning opportunities are accelerating. With these come challenges which are also increasing. Discomforts are shaking people out of their comfort zones. These are necessary on the path to ascension as they help us release attachment to the physical world. As the times become more challenging priorities change and many things that are now important will lose their importance as we become less inclined toward the material and more inclined toward the spiritual.

Suffering comes our way to help forge us as angels. Suffering helps us come closer to spirit and ascension. Suffering on the physical plane is unavoidable and acceptance of it with resignation helps us bear it. This is the lesson of the ages. If we let go to love, light and selfless service the suffering we have to endure will lessen. Saints and sages down through the ages have recognised the value in suffering. Neem Karoli Baba, a famous Indian sage and spiritual influence for many in the West, including Steve Jobs, Julia Roberts and Mark Zukerberg said, *"I love suffering; it brings me so close to God."*

Kahlil Gibran, famously captured the value of difficult experiences when in *The Prophet* he said:

> *"Is not the cup that holds your wine the very cup that was burned in the potter's oven?*
>
> *And is not the lute that soothes your spirit, the very wood that was hollowed with knives?*
>
> *When you are joyous, look deep into your heart and you shall find it is only that which has given you sorrow that is giving you joy."*

Christian Morgenstern (1871-1914) wrote:
Our desire no more to suffer causes only new pain,
Thus will you never shed your garment of sorrow,
You will have to wear it until the last thread,
Complaining only that it is not more enduring,
Quite naked must you finally become,
Because by the power of your spirit,
Must your earthly substance be destroyed,
Then naked go forward in only light enclosed,
To new places and times, to fresh burdens of pain,
Until through myriad changes a god so strong emerges,
That to the sphere's music you your own creation sings.

The best way to alleviate pain and suffering is to focus on alleviating the pain and suffering of others. If we treat others as we would wish to be treated and support them in their difficulties we will mitigate our own.

To quote **William Blake**:
Seek love in the pity of other's woe,
In the gentle relief of another's care,
In the darkness of night and the winter's snow,
In the naked and outcast, seek love there.

The golden key to liberation from suffering is the love found in compassion. If we give ourselves in selfless service and act out of unconditional love then misery is replaced by joy, even in the most dire situations. As we saw in war torn Syria, where children in a desperate situation found joy in looking after each other.

Diamonds are forged through exposure to intense heat and pressure. Similarly, history reveals that the human spirit grows strongest through adversity and difficulty and is weakened by comfort and leisure. Life on Earth is driven by fear but the existential process of wrestling with fear and eventually overcoming it is strikingly similar to the transformation of coal into diamonds. The journey through lives of fear and the ultimate triumph over fear is the evolutionary leap that brings a soul into self-mastery and ascension.

Buddha taught that suffering is caused by the frenzied chase between craving pleasure and avoiding suffering. He taught the solution to this pathology was to train the mind through meditation and to cultivate equanimity. Freedom comes from being free from bondage to pleasure or pain. Rudyard Kipling crystallised this perennial wisdom in his famous poem *If*: ...*If you can meet with triumph and disaster and treat those two imposters just the same.*

The spiritual path of appreciating the value of suffering is the complete opposite of the message put out in the materialistic civilisation of the twenty first century. The overarching cultural message is to pursue pleasure and avoid pain and suffering and that the source of ultimate value is the gratification of desire. Disastrous to ascension, the creed of materialism is: *Greed is good and money is God.*

Every human being will be offered the same opportunity to ascend but we are surrounded by every conceivable opportunity to become attached to the physical plane. A cacophony of activity is here to distract us from the purpose of our human life, which is to graduate and become an ascended master. There is purpose to this process of temptation. Those who are not attached to money and material things, who are dedicated to unconditional love and selfless service, they will make it through the door of light into the fifth dimension. Graduation is always preceded by examination. Everything pertaining to materialistic civilisation is an examination for ascension. If we judge good and evil we have missed the point. That is the original sin!

Saints and spiritual teachers have taught that the main purpose of human incarnation is to love and serve. Fortunate are those who can resist the lure of accumulating excessive money, possessions and material things, who can be happy without success or acclaim and have wisdom to avoid other forms of entrapment in today's world. Francis of Assisi and Buddha set the trend by turning their backs on material comforts for the sake of purity of heart and selfless service to humanity and to all life.

Masters of divine light, like Francis, Jesus, Gandhi and Buddha Gutama, have come into the world throughout history to offer a guiding hand. By example they showed that love and forgiveness, humility, poverty and service are far more valuable and ennobling than possessions, wealth, power and the esteem of others as most people are deluded by the false glitter of a world devoted to materialism

Despite the pressure for us to strive for material gain the world's sages warned us again and again not to fall into the trap of attachment to things in the physical world. Great saints like Yesua, Rumi, Kabir, Neem Karoli Baba, St Theresa of Avila and St Claire taught us that we are here to avoid materialism. We are here to awaken to our spiritual identities. They set the example of how to build up wealth in the heart where we can accumulate the real treasure that is imperishable. This is the imperative of ascension.

As ascension beckons the clarion call is sounding to live out of compassion and love. To fully participate in the world without attachment to it is the challenge of our day. Attachment to love alone is the answer. The time is now to prepare for our journey home, for our transition into eternity by being in the world but not of the world. This we can achieve by appreciating that the physical world is nothing but a passing phase in our total existence. In higher dimensions the Earth is called *The Planet of the Children* because it is a womb, nursery and *academy for angels*; a school for our angelic advancement and ascension. This is our opportunity to become mature galactic citizens. The time is now for us to win our wings and find our freedom to fly throughout the Universe.

CHAPTER 2
THE ACADEMY FOR ANGELS

Einstein's equation $E=mc^2$ equates mass with energy. In The Quantum Vortex I put forward a theory that the smallest particles in matter are spinning energy. That has enabled me to displace the idea of particles in the atom as bits of material substance and replace it with a vision of subatomic particles as vortices of energy. Imagine them as whirlpools of light.

In *Superenergy* I pressed home the point that we have been deluded to believe substantial material particles are the basis of our world. The non-substantial nature of energy reveals the world is more like a dream than a concrete reality. This is borne out in near death experiences where people report that they entered another the world which felt like home. They said it was as real than this world. When they were there the memory of this world faded away. Arriving there they felt it was as if they were waking up from a dream.

Down through the ages spiritual teachers have warned us of the delusional nature of the world we live in and quantum theorists back this up by revealing that the Universe is more like a thought than a material construct and consciousness is the bedrock of reality. We have been warned not to be taken in by the outward appearance of the world nor to prioritise the things of this world. We have been told that the masses are asleep as though in a dream. Many things they do are of little consequence to their true destiny. So what is our true destiny. I believe a clue is in a book by one of England's greatest cosmologists. In *The Intelligent Universe*[1] Sir Fred Hoyle, one time president of the Royal Society, wrote: *I don't like the idea of a single God creating the Universe. I prefer the idea in the Roman myths of gods acting as managers in an already existing Universe.*

The vision of the Universe I presented in *Superenergy* is in line with Hoyle's thinking. The redaction of *elohim* in the earliest Hebrew texts to *God* in the

1 Hoyle F. The Intelligent Universe, Michael Joseph, 1983

Christian Bible supports his view because elohim means *the gods* in Hebrew. God as depicted in the Old Testament appears more like the chair of a committee of universal managers than the creator of the Universe. In the Bible story of Abraham nearly sacrificing his son, *God* is clearly revealed as an angel. On reading the Bible if angels are universal management, God appears to be the senior amongst them.

If angels are gods responsible for the management of the Universe they would not be capable of doing so without education and training. One would expect the Universe to provide facilities for educating and training angels for managerial responsibility. It would seem obvious there would be academies for angels in the Universe. I am of the opinion the Earth is one.

As well as a womb where angels are gestated the Earth appears to be where angels are trained and suffering in the human condition indicates the training can be harsh. No other explanation for human suffering on the scale we witness would be reasonable. I wrote about it in the lines of my song, at the front of this book:

The plan of ages slowly unfolds,
Within its great purpose man evolves,
Ages of suffering grim to bear,
But how else would we learn to love, care and share,
So let not your problems hold you in their sway,
Nor let your troubles sweep you away,
These are just the trials that make us strong,
They'll keep on coming while to the Earth we belong.

My view is that the Earth is a *boot camp* for angels. This makes sense of beliefs held by the Gnostic sect of *Cathars.* Cathar pastors were called *Perfecti.* They taught that someone called *Satanas* made Man in his own likeness and ordered angels to enter his human creations. *Les Questions de Jean*, is a record of an Inquisition interrogation of a Cathar Perfecti who said, *"And he Satan imagined in order to make man for his service, and took the lime of the earth and made man in his resemblance. And he ordered the angel of the second heaven to enter the body of lime; and he took another part and made another body in the form of woman, and he ordered the angel of the first heaven to enter therein. The angels cried exceedingly on seeing themselves covered in distinct forms by this mortal envelopment."* (*The Gnostics* p.74) [2]

2 Churton T, The Gnostics, Weidenfeld & Nicolson, 1987

This tract suggests that angelic spirits were coerced into incarnation on Earth by an entity called *Satan.* In *The God Hypothesis,* Dr J. Lewels, spoke of the Mandaens, who believed in a universe, divided equally into the worlds of light and darkness. *"To them, the physical world, including the Earth, was created and ruled over by the Lord of Darkness...variously called Snake, Dragon, Monster and Gian...thought to be creator of humanity."* [3]

Tobias Churton from The Gnostics: [2] which accompanied a television series by the same name on BBC, TV wrote: *There were Jewish schools, much given to speculation on the nature of God and the constituent beings which constituted his emanation or projection of being. Some of them appear to have been profoundly disappointed with the God of the Old Testament and wrote commentaries on the Jewish scriptures, asserting that the God described there was a lower being, who had tried to blind Man from seeing his true nature and destiny. We hear their echoes in some books of the Nag Hammadi Library, namely, 'The Apocryphon of John' and 'The Apocalypse of Adam. They believed in a figure, the 'Eternal Man' or 'Adam Kadmon' who was a glorious reflection of the true God and who had been duped into an involvement with the lower creation, with earthly matter, ruled by an inferior deity who, with his angels, made human bodies.*

William Blake also perceived the Old Testament god, Jehovah as having fallen from an original state of virtue and dignity as did Yeats:...*He is recognizably the 'Elohim' of the Old Testament in his 'aged ignorance', carrying the books of the Law, and yet Blake's Creator is not wholly evil; for as the 'Ancient of Days' on the frontispiece of 'Europe' (1793) we see the Creator with his golden compasses, who, though in part fallen, 'derived his birth from the Supreme God; this being fell, by degrees, from his native virtue and his primitive dignity'* W. B. Yeats (The Gnostics, p.147)

In the Gospel of John, Jesus declared the Old Testament god to be Satan, otherwise known as the *Devil.* He vented at the Jews *"You belong to your father, the devil, and the desires of your father you want to do. He was a murderer from the beginning, and does not stand in the truth, because there is no truth in him. When he speaks a lie, he speaks from his own resources, for he is a liar and the father of it."* (John 8:44)

So who is this devil called Satan, intimated even by Jesus to be the Old Testament god? We must delve in the Hindu scriptures to find an answer to that.

I believe Satan is *Sanat Kumara.* In Hinduism he is revered as the eldest son of Brahma. In Vedic scriptures a Kumara is a prince and Sanat is also known as

3 Lewels J., The God Hypothesis, Wildflower Press, 1997

Skanda, son of Shiva. Portrayed as ever youthful Sanat Kumara is considered to be one of the progenitors of mankind. According to a number of sources, including Mark Prophet, Sanat Kumara is known as the *Ancient of Days.* If that is true then by commonality of appellation we see a link between Sanat Kumara and Jehovah in the Bible. In a vision, Daniel (Daniel 7:9) saw the *Ancient of Days* seated on the 'throne' as the God of his fathers Abraham, Isaac and Jacob. It was the Ancient of Days who appeared to Moses in the burning bush and declared himself as YHWH, sounded by some as *Yahweh*, or *Jahveh*, commonly pronounced *Jehovah.* Sanat Kumara is also known as the *Planetary Logos,* which means he is the *god of this world.* Many, including the Cathars, Mandaeans and Christians, consider the 'God of this world' to be Satan. So what is the link between Sanat and Satan? If you look at the two names side by side you will see one is an anagram of the other.

Sanat Kumara is not evil. In the role of Satan, or the Devil, he took the hard task. His position as serving a higher authority is made clear in the Bible. In the Book of Job he attended meetings and was sanctioned by a higher authority to test Job and make his life difficult for him. In his role as the tester and the tempter Satan behaves in a way that can be perceived as evil but the evil would not be his if it were authorised by a higher authority. Satan may have been a *father of lies* because he was working to a hidden agenda. I believe it was in his role to keep us in the dark. The second century Gnostic, Valentius, spoke of the Old Testament God as a demiurge - a lesser authority. Valentius also said he was: *...determined his creatures (us) shall remain unaware of their source.* (The Gnostics, p.55)

I believe Sanat Kumara was the leader of the Elohim. I believe he was responsible for the redaction of the gods to a single God. He had an agenda and he needed a chosen people to achieve it. They had to obey his command and worship him alone. This is abundantly clear in the first of the ten commandments, *"I am the Lord thy God, which brought thee out of the land of Egypt, from the house of bondage. Thou shalt have none other gods before me.* (Deuteronomy 5:6-7)

I also believe Sanat Kumara was Enlil, the senior god in the ancient Sumerian texts. Abraham was Sumerian and Enlil was the main Sumerian god. It would make sense if the god of Abraham who returned to Moses was Enlil.

Enlil wasn't the bioengineer of humankind. That role fell to his brother Enki but Enlil was in the position of overall command of *Project Earth* therefore he would have been 'responsible' for our origins. The Sumerians said that Enlil and Enki answered to a higher authority called *Anu* and in the Sumerian records they were reptile in origin. [4]

As the pieces of the jigsaw puzzle came together, Enlil appeared to correspond to Archangel Lucifer in the Hebrew texts and Enki to Archangel Michael. Enlil was Jehovah and Enki incarnated as Jesus. That would make sense as it is only natural that the one who made us should be our saviour. If Jesus was a reincarnation of Enki his outburst in John 8:44 is understandable because there was enmity between Enki and Enlil and Enlil wasn't nice. In the garden of Eden when one of the reptilian team told Eve the truth about her origins he was treated harshly: *And the Lord God said unto the serpent, "Because thou has done this, thou art cursed above all cattle, and above every beast of the field; upon thy belly thou shalt go, and the dust thou shalt eat all the days of thy life.* (Genesis 3: 14)

According to Hindu scripts, Brahma sent his four sons, Sanat, Sananda, Sanatana and Sanaka into the first galaxy to create life. In the Hebrew tradition these would correspond to Lucifer, Michael, Gabriel and Raphael respectively. According to Hindu teachings the princes or kumaras had to overcome extremes of difficulty, challenge and adversity to achieve their original mission to establish life in the very first galaxy and as a result they evolved spiritually to a level that was never surpassed. In a phone call with the author, Stewart Wilson, Stewart said the Kumaras were charged by the highest authority to bring the angelic hordes, behaving much as depicted in Greek and Roman mythology, up to their standard. Enter planet Earth.

It is common for star systems to follow orbits and according to Peter Deunov the solar system we inhabit has been passing through a contaminated area of space for thousands of years. It is possible the contaminated zone is in the orbit of our solar system. If that were so it might pass between 'light and dark', in and out of this zone of density, on a recurrent basis. The entire orbit lasts for approximately 26,000 years. A part of this may be spent in the contaminated zone and a part in uncontaminated space. Perhaps because of this astronomical situation, and its water, the Earth was chosen by the Kumaras as a womb and academy for angels.

Like many educational organisations, which progress from elementary to advanced teaching, the terrestrial institution for angelic advancement appears to follow four phases of education; the equivalent of nursey school, primary school, secondary school and university. These depict the four ages of

---

4 Boulay R.A. Flying Serpents and Dragons: The Story of Mankind's Reptilian Past, Book Tree, 1997 (revised 1999)

humankind recognised in a number of cultures. In the West they were known as that. In the East they were known as the four yugas beginning with the Sat Yuga, a golden age of light, unity and truth and ending with the Kali Yuga, an age of brutality, darkness and discord.

The four ages of humankind were described in the mythologies of ancient Greece and Rome. To quote Thomas Bulfinch:[5] *The first age was an age of innocence and happiness called the Golden Age. Truth and right prevailed though not enforced by law, nor was there any magistrate to threaten or punish. The forest had not yet been robbed of its trees to furnish timbers for vessels nor had men built fortifications round their towns. There were no such things as swords, spears or helmets. The Earth brought forth all things necessary for man, without his labour in ploughing or sowing...*

Then came the Silver Age, inferior than the Golden but better than the Brass... For the first time men had to endure the extremes of heat and cold, and houses became necessary. Caves were the first dwellings, and leafy coverts of the woods, and huts woven of twigs. Crops would no longer grow without planting. The farmer was obliged to sow the seed with a toiling ox to draw the plough.

Next came the Brass Age, more savage of temper and readier to the strife of arms, yet not altogether wicked. The hardest and worst was the Iron Age. Crime burst in like a flood; modesty, truth and honour fled. In their places came fraud and cunning, violence, and the wicked love of gain. Then seamen spread sails to the wind, and trees were torn from the mountains to serve for keels of ships, and vex the face of the ocean. The Earth, which until now had been cultivated in common, began to be divided off into possessions. Men were not satisfied with what the surface produced, but must dig into its bowels, and draw forth from thence the ores of metals. Mischievous iron and more mischievous gold were produced. War sprang up, using both as weapons; the guest was not safe in his friend's house; and sons-in-law and fathers-in-law, brothers and sisters, husbands and wives could not trust one another. Sons wished their fathers dead, that they might come to the inheritance; family love lay prostrate. The Earth was wet with slaughter..."

The legend of a progress in ages from a heavenly paradise of innocence and kindness, in community with the gods, followed by descent into hardship, brutality and despair with separation from the gods and loss of our divine heritage is the mythological cornerstone, not only of Greek and Roman civilisation but of cultures throughout the world. These stories, demeaned as mere

5 Bulfinch T. The Golden Age of Myth & Legend, Wordsworth Reference, 1993

myths, depict the learning cycle for humanity. The idea in Western religion that this is a progress from good to evil is naïve.

In the Hindu religion the four yugas of humankind are described in detail in the *Vedic* scriptures as four periods of growth for the souls incarnating on Earth. They also describe progress from an age of gold to an age of iron, the equivalent to the educational phases from nursery to university, as a necessary process in the evolution of humanity. The education of angels appears to be a progression from comfort and joy to discomfort and misery. The progress is toward ever increasing difficulty, challenge and adversity. This mimics the Kumara's experience on their original mission in the very first galaxy, which led them to become unsurpassed in their managerial abilities in the Universe. Their agenda has been to duplicate the arduous situations they went through. Because it involves immense pain and suffering they have had to hide the curriculum from their angelic pupils until the end of the course. Angels enjoy free will so they cannot be forced against their will. They had to be enticed and tricked into attending Satan's academy. Satan could only order angels into incarnation if he had already tricked them into following him, much like soldiers. Recruits are first tempted and gloried into signing up then when they are soldiers they have to obey orders.

Everything that happens on Earth is part of a curriculum designed to build strength of spirit. Some angels voluntarily descended from a heavenly state to the gravity plane of the Earth to serve as teachers and learn more lessons. Some angels have fallen to Earth to be reformed and some were gestated here and are being put through their paces. No matter how they came, the *Earth Academy of Angels* acts as a crucible of fire where dross is separated from gold and spiritual coals are transformed into diamonds through the heat and pressure of bitterness, cruelty and brutality.

According to ancient records each age appears to have been separated by a cataclysm[6] most probably caused by a catastrophic pole shift.[7] Priests and prophets of doom proclaimed they were vengeance wreaked upon a sinful humanity by wrathful gods. It is more likely recurrent disasters that afflict the Earth are inevitable Earth changes[7] that are in no way linked to human behaviour. I consider the geophysical events serve to clear classrooms at the end of one yuga in preparation for the next in the progress from Sat Yuga to Kali Yuga. Neither is the Kali Yuga or Age of Iron bad for us. We have to

6 Velikovsky I. Worlds in Collision, V. Gollancz, 1950

7 Hapgood C., Earth's Shifting Crust: A Key to Some Basic Problems of Earth Science, Pantheon Books, 1958

rise above judgement of good and evil to appreciate the big picture of what is going on in the Universe. Especially it is not for us to judge Sanat alias Satan.

The Kali Yuga represents the age when evil and ignorance are rampant on the Earth but we need to experience evil as well as good. The greed and selfishness, materialism and carnage in this last age of the yuga cycle is a vital part of our angelic education. Satan's role is to lead us into temptation in order to reveal our true colours.

Some people are anxious to learn about a group of supposedly deviant celestial beings called *archons*. The archons are referred to in Gnostic teachings discovered in Nag Hammadi, Egypt, in 1947. The teachings date back hundreds of years BCE. The Gnostics believed that archons maintain the Earth as a prison for human souls. 'Archon' is the Greek equivalent of kumara meaning prince or ruler. The Gnostic treatises tell us that the archons serve a *demiurge,* an intermediary celestial entity that stands between the human race and the transcendent God. Gnostics believe that only through Gnosis, a direct revelation of the Knowledge of the soul's divine nature, can a soul be liberated from enmeshment in the material dimension and return to the upper worlds of the divine.

The Gnostics say archons operate, under Satan, from a parallel world. I would interpret this as the archons operating from the hyperphysical fourth dimension. In the Academy for Angels, they are equivalent to teachers in a boarding school. Their role is to strengthen our spirits through tests, trials and ongoing adversity. Archons are adversarial forces that train us through suffering to foster our spiritual constitution. They sometimes incarnate for this purpose and rise to positions of psychopathic leadership in human society.

In the Book of Enoch[8] archons are called *The Watchers*. They are said to be fallen angels. It is they who judge us. They set up temptations and encourage addictions and attachments to enable us to learn vital lessons on the physical plane. As our teachers they evaluate our progress.

The job of teachers in a boarding school is also to stop pupils absconding. An additional role of the archons is to prevent juvenile angels from escaping the Earth boot camp for angels. As with school-masters they use whatever measures are necessary to keep pupils within the bounds of the academy until the course is complete.

Teachers of a sadistic nature may be vilified by pupils as monsters and devils

8 Schnieders P.C. The Books of Enoch, International Alliance Pro-Publishing, 2012

and the Gnostics regarded the archons as 'rulers of darkness'. We must rise above that judgement.

Sanat, alias Satan, is leader of the archons. He is the equivalent of headmaster in the boarding school analogy. As well as being the Ancient of Days - the oldest spirit - he is also known as *Gatekeeper of the Beyond the Beyond.* Beyond the physical third dimension is the hyperphysical fourth dimension so 'beyond the beyond' is the superphysical fifth dimension. Satan's role is not only to oversee the archon programme of angelic education, tempting and testing us to bring out to the fore our inner hidden tendencies, he is also charged to guard access to the fifth dimension. With the help of the archons he must ensure we don't break out into the Universe at large until we have passed out of quarantine by the most stringent test, which is the ability to be unconditional in our love and forgiving in our attitude to others. For that we need wisdom and understanding. We need to see the bigger picture.

As humans we have the potential to be planetary pathogens. We may despise Satan but the Universe relies on him to maintain galactic health and safety. The God of the Old Testament, head of the archons, is not evil. He is fulfiling a role to safeguard other populations in the Universe

Adversarial archons incarnate and rise to positions of power and promote materialism to keep us in ignorance and to tempt our propensity to attachment. They do this to test our judgments and bring out in us our inner tendencies.

Our teachers are not all adversarial angels of the dark. Some are angels of the light. Together with the archons they work to help us develop our inner strength and ability to overcome difficulties and resistance, obstacles and limitations and to help us conquer apathy and spiritual inertia.

We are portions of the *One Cosmic Being* trapped in the cage of an impure and ignorant mind and our teachers mirror and reflect this. Demonised as Pluto, Hades, Ahriman and Lucifer with his fallen angels, they reflect the dark side of humanity. Swedenborg taught that the states of heaven and hell exist within us. We need look no further than ourselves to see a universe of angels and the devils.

Everything happens for a purpose. There are no mistakes. According to Gnostic thought, *Sophia* broke away from the spiritual planes to form the physical realm of the Earth and the hyperphysical anima mundi, the soul of the Earth. As wisdom embodiment of Gaia, she was creating a womb; the *Earth Womb of Angels.* She was also setting up the *Earth Academy for Angels.* She did not cause our fall; she was building our school. Sophia is our divine mother but also the equivalent of school governess in the boarding school analogy. Gestating us and overseeing our education as up and coming angels, through trial and tribulation, she is ultimately responsible for forging us as gods.

In Hindu mythology *Adi Para Shakti* is the Supreme Mother who gave birth to the trinity of Brahma-Vishnu-Shiva. In the yogic tradition an aspect of Adi Para Shakti, the divine feminine origin of life and all that is, broke away from perfection to set up our world. Shakti in Hindu mythology is the exact equivalent of Sophia in the Greek mythology. Similar wisdom teachings come through from ancient times in diverse mythologies. Mythologies are teaching stories. Hidden within the legends, myths and stories is a current of esoteric knowledge and wisdom preserved over the centuries for sincere seekers of truth.

Everyone has a purpose and everything has a place in the plan of ages to teach us and put us through our training. Fallen angels that use their powers to dominate us serve to educate us about the dominating tendency in our own nature. Psychopaths, despots and tyrants in our families or people we come up against at work may increase our suffering but it is for our ultimate good. There is no absolute evil in them. If they are in our lives they are mirrors for us. If we see evil in others it is a reflection of evil in ourselves. If we see good in others that is a reflection of our goodness.

Gems are polished by tumbling with grit and gold is purified by fire. We are forged through anguish by forces perceived as evil for the express purpose that we come into a perfection of our own making. We may complain but as Rumi said, *"If you are irritated by every rub, how will you be polished?"*

The plan of ages unfolding from gold to iron, goodness to evil and back again, finds symbolic representation in the Yin Yang symbol. The Chinese principle of Yin and Yang is that the greatest good carries the seed of evil and the greatest evil carries the seed of good and when each polarity comes to its fullness it flips to its opposite.

Most of us were seeded into our cycle of human incarnation at the beginning of a golden age some twenty six thousand years ago. The golden age was sweet and harmonious for us as angel newborns. We found ourselves in the equivalent of pre-school. Ascended humans from previous cycles returned to the planet as gods or divines, 'the shining ones,' to conceive, birth and gently guide us into the terrestrial system of angelic education. From there we moved on to a silver age, the equivalent of primary school, then onto a brass age, the equivalent of secondary or high school, until eventually we entered the iron age, or Kali Yuga, the equivalent of university where we have been learning to overcome every imaginable humiliation, fear and terror until we know only love. If we learn the lesson that love prevails then passing out of quarantine, through ascension, we will have access to the Universe at large. Some of us will choose to return to Earth, after its purification and regeneration to seed the next generation of angels.

Fortunate are those who pass the tests at the end of the course and graduate in ascension. We have come a long way to reach the threshold when the Earth leaves the zone of density. That is when the intrinsic speed of energy in every subatomic particle and quantum of energy in the atoms of Earth will revert to the speed of superphysical energy. This occurs as we pass out of the contaminated space that is depressing the speed of light.

Most humans have hearts of gold. Most people are full of love and light and are ready to ascend with the Earth but the Earth is dominated by corrupt corporations, governments and people in power serving self-motivated interests. The Earth is also plagued by parasites and polluted with plastic and contaminated with chemicals. And then there are people who are not yet ready to participate in the ascension, nor would they want to. The planet must be cleansed before it leaves the zone where darkness and density prevail. There has to be a separation between those who feel ready to proceed into the fifth dimension and those who prefer to live on in density.

An Indian epic, the *Mahabharata*, records a previous event that fits the ascension predictions precisely. This supports the idea the Earth may have passed through previous 26,000 year cycles of four, 6,500 year ages or yugas.

From his extensive study of ascension, the research journalist, William Henry, says ascension will precede a solar event called a *Samvartaka*. To quote him, *"In the Indian epic poem the Mahabharata ('Great Tale'), which some consider the most incredible tale ever told...we learn of a cyclical solar event called the Samvartaka Fire...the Zoroastrians called it the Frashokereti, meaning the final renovation or 'refreshing' of the universe, when evil is vanquished and all will be in perfect unity with God and a new Saviour emerges...The Mahabharata tell us the Samvartaka Fires emanate from Surya, the Sun, and are characterized as a cloud filled with 'wreaths of lightning'. Strange clouds accompany the lightning. It is wonderful to behold... It looks exactly like NASA images of coronal mass ejections from the Sun! It seems pretty obvious that the Ascension Fire of Christ comes through the Sun... One aspect of the contemporary interpretation of the Samvartaka fire that most interests me is...that masses of souls were spontaneously lifted off the earth during previous Flash Events and that celestial intermediaries, angels, whose bodies resembled the Sun, assisted in this mass ascension... The Mahabharata says that during this 'End Time' event, luminous, humanoid beings will be seen to appear on earth. [David] Wilcock says...the luminous ones of the Mahabharata are the Seraphim beings of Judeo-Christian tradition".* [9]

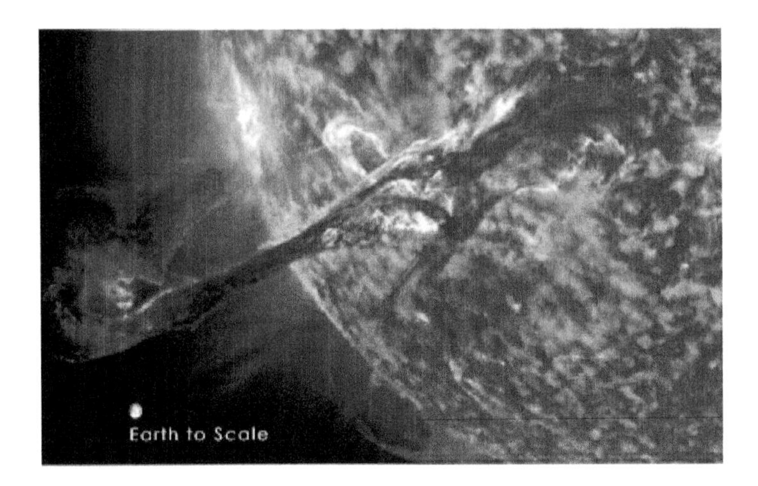

The appearance of an army of seraphic angels would be tantamount an angelic invasion of the Earth. Evil invade; Good invites. The predicted arrival of Seraphim cannot be an invasion, it has to be a first wave of graduates from the Academy of Angels returning to usher the rest of humanity to graduation through the invitation to ascension.

The Ascension Plan is to invite a number of people to ascend and then they return to Earth in the ascended state to lead others to ascension. This will avoid an affront to human rights by an invasion of the light. People on Earth who have prepared themselves will be the first to be ascended to become luminous human angels. Then, returning to Earth to help everyone else they will bring in the light from among the human race not from outside. Jesus appeared to Francis of Assisi as a seraph. If that is what he became on his ascension it follows he would call on us to ascend likewise and follow him into the ranks of the Seraphim.

Angels and archangels occupy lower ranks in the divine order. The Seraphim are at the highest levels of angelic divinity. Amongst the most powerful angels in the Universe, these Guardians of Galaxies are closest to the Source. Angels become Seraphim by going through the arduous evolutionary process in the challenging cycles of human incarnation. The difficulties, challenges and adversity, the humbling and the suffering along with the sheer richness of the human experience, physical, emotional, mental and spiritual, available on

9 https://www.williamhenry.net/2018/04/ascension-fire-the-solar-flash-and-the-return-of-christ/

Earth, are unsurpassed in angelic evolution anywhere.

Like the Samurai sword that is forged through a repeated cycle of heat and hammering until it is above and beyond any other, so angels and archangels hammered, heated and humbled, could become Seraphim through everything they have had to endure by being human. Happiness and suffering both have their place. Most people seek happiness and avoid suffering but it is through the endurance of suffering that we make the greatest advancement. All human suffering will have been worthwhile if it is the path that leads to the highest levels of the Universe. This amazing opportunity could be available to everyone who graduates with honours from the Academy of Angels.

CHAPTER 3
THE EARTH CHANGES

In the *Earth's Shifting Crust*[1] Earth Scientist Charles Hapgood presented compelling evidence that the crust of the Earth periodically moves over the main body of the planet. Hapgood pointed out that major Earth changes occur cyclically in a period of less than ten thousand years.

In his book of seismic implications, the author rewinds the earth clock some ten thousand years. The geological evidence shows that rivers were flowing on the continent of Antarctica less than ten thousand years ago while the northern States of America were covered under a sheet of ice two miles deep. The pole was at Hudson Bay. Less than ten thousand years ago something happened that caused America to be displaced some two thousand miles towards the equator, shifting the pole from Hudson Bay to the Arctic Ocean.

What is even more difficult to grasp is that the pole under Hudson Bay was a South Pole. Whatever happened less than ten thousand years ago also caused a reversal in the Earth's poles. Earth sciences reveal that every few thousand years disparate regions of the Earth are visited by radically different climatic conditions from polar ice to desert sands, from equatorial rainforest to temperate woodland. It's difficult to believe that the Lake District of Northern England was once covered by a glacier, while just a few hundred miles to the south the red sand cliffs of East Devon were once desert dunes and on the Jurassic coast just a few miles further on, in the county of Dorset, there is evidence of dinosaur bones. The implication of this is staggering. These cold blooded reptiles lived in hot climates. They could only survive in tropical regions. All the evidence points to dramatic climatic changes occurring periodically all over the Earth, which can be explained by the crust shift hypothesis.

---

1 Hapgood, C., Earth's Shifting Crust: A Key to Some Basic Problems of Earth Science, Pantheon Books, 1958

Because the Earth sciences show that pole shifts are recurrent as we move in time further from the last event we inch ever closer to the next one coming. That is why prophets and messiahs, spiritual teachers and enlightened scientists warn us to be prepared for an event which could destroy civilisation and plunge survivors into a Stone Age.

Albert Einstein was impressed by Hapgood's manuscript and wrote a foreword to his book.[1] In Einstein's words: *I frequently receive communications from people who wish to consult me concerning their unpublished ideas. It goes without saying that these ideas are very seldom possessed of scientific validity. The very first communication, however, that I received from Mr. Hapgood electrified me. His idea is original, of great simplicity, and – if it continues to prove itself – of great importance to everything that is related to the history of the earth's surface.*

A great many empirical data indicate that at each point on the earth's surface that has been carefully studied, many climatic changes have taken place quite suddenly. This, according to Hapgood, is explicable if the virtually rigid outer crust of the earth undergoes from time to time, extensive displacement over the viscous, plastic, possibly fluid inner layers. Such displacements may take place as the consequence of comparatively slight forces exerted on the crust, derived from the earth's momentum of rotation, which in turn will tend to alter the axis of rotation of the earth's crust.

In a polar region there is continual deposition of ice, which is not symmetrically distributed about the pole. The earth's rotation acts on these asymmetrically deposited masses, and produces centrifugal momentum that is transmitted to the rigid crust of the earth. The constantly increasing centrifugal momentum produced in this way will, when it has reached a certain point, produce a movement of the earth's crust over the rest of the earth's body, and this will displace the polar regions toward the equator.

Without a doubt the earth's crust is strong enough not to give way proportionately as the ice is deposited. The only doubtful assumption is that the earth's crust can be moved easily over the inner layers…

Albert Einstein pointed out that for Hapgood's account to be plausible an explanation for heat in the Earth, sufficient to enable such massive periodic events to occur, had to be found. For the crust to slip the heat in the Earth would have to be sufficient to turn the *magma* - molten rock - under the crust from the consistency of glue to a lubricant - a bit like turning sticky honey to runny honey when it is heated. Only that would enable the crust to slide freely over the inner layers. The Charles Hapgood hypothesis requires that there must be an enormous source of heat in the Earth, which is released periodically. Something has to be going on in the Earth to enable regular Earth changes of such a magnitude as to enable the crust to roll over the poles not just once but again and again.

In *Earth's Shifting Crust,*[1] Hapgood admits that the heat in the Earth is a mystery. In fact it is one of the main unsolved problems in geology. He quotes from *The Internal Constitution of the Earth*[2] by Beno Gutenberg: *A vast amount of research has been devoted to this subject, but the fact remains that the origin and maintenance of the earth's internal heat continues to be one of the outstanding unsolved problems of science.*

The account for gravity in the quantum vortex theory could explain the periodic rise of heat from the core of the Earth sufficient to reduce the viscosity of the molten rock under the crust so that instead of acting as glue, gripping the crust to the inner layers of the Earth, it acts as a lubricant enabling the crust to slide freely over them. That could explain periodic pole reversals. Global warming may also be linked to the rise of heat from the centre of the Earth.

As I developed the physics of the quantum vortex I realised there may be a sub-stellar 'mini' black hole at the Earth's core generating a vast amount of heat from the annihilation of matter and antimatter. This could be rising from the core and account for the heat in the Earth. Matter-antimatter annihilation is the greatest known source of energy in the Universe. It would more than suffice to generate the heat necessary to turn molten rock - magma - from the consistency of sticky glue to that of slippery oil.

Because I have a unique awareness that global warming could be linked to the periodic shift of the Earth's crust, I take the ascension message very seriously. I feel a sense of urgency to warn of what may be happening right under our feet. This is not to spread fear but to encourage as many people as possible to rethink their priorities.

Stories, epic poems and ancient records from history tell of great civilisations disappearing, without a trace, as a result of periodic global catastrophes.

---

2 Gutenberg B., Internal Constitution of the Earth, Dover Books, 1951

Legends of the lost civilisations of Lemuria, Mu and Atlantis, may be such ancient memories. But if heat rises from the centre of the Earth, due to the annihilation of matter and anti-matter, why would it be periodic?

In *The Quantum Vortex* I suggest that beyond the centre of the Earth there is an antimatter Earth as antimatter exists beyond the centre of everything. Annihilation occurs where matter meets antimatter. This could be happening at the core of the Earth. To explain the release of annihilation energy from the centre of the Earth I lent on Steven Hawkins's *sub stellar mass black hole*,[3] theory to propose there might be a mini black hole at the core of the Earth.

If I am right, to begin with the annihilation energy would be trapped in the black hole and then periodically it would be suddenly released. This is because of the way black holes behave. In *The Quantum Vortex* I explain how the attraction between matter and antimatter, through the centre of black holes, causes extreme gravity. Then I accounted for gamma ray bursts from distant galaxies by postulating that black holes behave as *geysers*.

If the heat in the Earth were coming from the annihilation of matter and antimatter, it would be released continuously from the singularity point at the centre of a black hole. To begin with the energy would be gripped by the intense gravity of the black hole. But then when the build-up of annihilation energy is sufficient to overcome the ability of the black hole gravity to hold it some of the trapped energy would escape to rise through the dense core of the Earth. The release of heat would not be immediate because the gamma ray energy, produced by the annihilation, would have to first be transferred into heat as it is absorbed by the dense iron core of the Earth. After building up to super-saturation levels a wave of heat would be released and then begin its convection through the thousands of miles of the inner layers of the Earth before reaching the magma under the crust.

After release of surplus energy the gravity of the mini black hole at the centre of the Earth would 'snap back tight' on the annihilation energy. The core would then cool down ready to start the cycle all over again. Meanwhile the wave of heat would be rising, inexorably, toward the crust of the Earth destined to clear human civilisation off it.

Geysers release energy periodically. The periodic release of annihilation energy by a *gravity geyser* could explain the global catastrophes that happen periodically to transform the face of the Earth and wipe out civilisations without trace and why this keeps happening. The gravity geyser could account for the

---

3 Hawking S. Gravitational collapsed objects of very low mass, Monthly Notices of the Royal Astronomical Society, 1971

climate changes, periodic pole shifts, mountain formations, continents rising and falling, great inundations and periods of global warming followed by ice ages that constantly afflict our planet. But why would these be characterised by the crust rolling over the poles? The 23.5 degree tilt of the Earth on its axis results in one pole being closer to the sun than the other at the solstice. The accumulating ice on the poles means that year on year the differential in the gravitational pull from the sun on the poles during a solstice grows stronger. Eventually, during a time of global warming the floating crust could slide over the poles. It would, quite literally, roll 'head over heels'. If the crust were to reverse its position over the poles, for people left on the planet after the cataclysm the sun would appear to reverse its direction in the sky. If there were a pole reversal now, after the event, the sun would appear to rise in the West and set in the East. Apparently such reversals has occurred many times in the past.

CHAPTER 4
THE POLE REVERSALS

In *Worlds in Collision*[1], Immanuel Velikovsky presents a number of extraordinary accounts from ancient history that reveal a similar story of repeated global catastrophes in the folk memory of practically every culture on Earth. If the ancient records he unearthed are to be believed the world would appear to have been through repeated disasters and each time the sun appears to have reversed its direction in the sky.

Velikovsky discovered in Herodotus' *second book of history*, reference to a conversation with Egyptian priests in which they asserted that since Egypt became a kingdom: ...four times in this period the sun rose contrary to his wont; twice he rose where he now sets, and twice he set where he now rises.[2]

A Latin author of the first century, Pomponius Mela, commented: "The Egyptians pride themselves on being the most ancient people in the world. In their authentic annals one may read that the course of the stars has changed direction four times and the sun has set twice in that part of the sky where it rises today."[3]

The Papyrus Harris records a cosmic upheaval when: ...the south becomes north and the Earth turns over.[4]

The Papyrus Ipuwer stated: The land turned round as does a potter's wheel and the Earth turned upside down.[5]

In the Ermitage papyrus reference is made to: ...a catastrophe that turned the

---

1 Velikovsky I., Worlds in Collision, V. Gollancz Ltd., 1950

2 Herodotus, Bk ii 142 (transl. A.D. Godley 1921)

3 Mela Pomponius, De Situ Orbis. i. 9. 8.

4 Lange H., Der Magische Papyrus Harris Danske Videnskabernes Selskab, (p58) 1927

5 Lange H., German translation of Papyrus Ipuwer 2:8, (pp 601-610) Sitzungsberichte Preuss, Wissenschaften 1903

Earth upside down.[6]

The sun was known in ancient Egypt as Harakhte. There can be no doubt that the catastrophe spoken of was associated with a reversal in the direction of the sun because elsewhere in the Ermitage papyrus there is reference to the original direction of sunrise: "Harakhte he riseth in the West."[7]

Texts found in the pyramids say the luminary ceased to live in the occident and shines anew in the orient.[8] Velikovsky went on to claim in the tomb of Senmut, architect of Queen Hatshepsut, there is a panel on the ceiling which shows the celestial sphere with the signs of the zodiac and other constellations in a reversed orientation of the southern sky where north is exchanged for south and east for west.[9] The simplest conclusion to draw from this anomaly is that this was how the night sky originally appeared.

In Politicus Plato wrote: I mean the change in the rising and setting of the sun and the other heavenly bodies, how in those times they used to set in the quarter where they now rise, and used to rise where they now set.[10]

The reversal of the sun in the sky was never a peaceful event. In Politicus Plato concluded: There was at that time great destruction of animals in general and only a small part of the human race survived.[10]

Velikovsky found in the drama Thyestes, by Seneca, a powerful description of what happened when the sun turned backward in the morning sky[11] which, in Timaeus, Plato detailed: "...a tempest of winds...alien fire...immense flood which foamed in and streamed out...the terrestrial globe engages is all motions, forwards and backwards and again to right and to left and upwards and downwards, wandering every way in all the six directions."[12]

In the sacred Hindu book 'Bhagavata Purana' four ages or Yugas are described. Each ends in a cataclysm in which mankind is almost destroyed by fire and flood, earthquake and storm.[13] These age changing events are documented by almost every culture in the world, from times when they were totally disconnected. Hesiod, one of the earliest Greek authors, wrote about four ages

---

6 Gardiner, Journal of Egyptian Archeology I, (1914); Cambridge Ancient History I, 346

7 Breasted, Ancient Records of Egypt III, Sec

8 Speelers L., Les Texts des Pyramides I, 1923

9 Pogo A., The Astronomical Ceiling Decoration in the Tomb of Senmu, Isis (p.306) 1930

10 Plato, Politicus (transl. H.N.Fowler, pp 49 & 53, 1925)

11 Seneca, Thyestes II. (transl. F.J.Miller 794 ff)

12 Plato, Timaeus, (transl. Bury, 1929)

13 Moor E., The Hindu Pantheon 1810

and four generations of men destroyed by the wrath of the planetary gods[14]. He described the end of an age as: The life giving Earth crashed around in burning...all the land seething and the oceans...it seemed as if Earth and wide heaven above came together; for such a mighty crash would have arisen if Earth were being hurled to ruin, and heaven from on high were hurling her down."[15]

The Persian prophet Zarathustra spoke of: ...signs, wonders and perplexity which are manifest on the Earth at the end of each age.[16]

The Chinese call the perish of ages Kis and number ten Kis from the beginning of their known world until Confucius.[17] In the ancient Chinese encyclopedia, 'Sing-li-ta-tsiuen-chou' the general convulsions of nature are discussed. Because of the periodicity of these convulsions the Chinese regard the span of time between two catastrophes as a Great Year. As during a year so during a world age the cosmic mechanism winds itself up and:...in a general convulsion of nature, the sea is carried out of its bed, mountains spring out of the ground, rivers change their course, human beings and everything is ruined, and the ancient traces effaced.[18]

An ancient tradition of world ages ending in catastrophe is persistent in the Americas amongst the Incas[19], the Aztecs and the Maya.[20] The Maya have a Long Count between global catastrophes similar to the Chinese Great Year.

In her book on global catastrophes in ancient times,[21] Lucy Wyatt suggested a date for Noah's Deluge seventy nine years after the beginning of the Long Count that ended in 2012. Her research indicated that cataclysms occur close to the target date of a Long Count or a Great Year but not on it. That would fit with the principle in science of a margin of errors. In a cycle of global cataclysms that are millennia apart one would expect a cataclysmic event to occur sometime within a margin of a few decades of a predicted date.

The Maya said each age ends with earthquakes at the solstice. The Maya reference to a worldwide catastrophe of global earthquakes at the solstice supports the Hapgood model of crust slips being related to the accumulation of ice on

14 Hesiod, Works and Days (transl H. Evelyn-White 1914)

15 Hesiod, Theogony (transl. H. Evelyn-White 1914)

16 . Muller:, Pahlavi Texts: The Sacred Books of the East, 1880

17 Murray. Historical & Descriptive Account of China, 1836

18 Schlegel G., Uranographie Chinoise, Wou-foung, 1875

19 Schlegel G., Uranographie Chinoise, Wou-foung, 1875

20 Humbolt A von, Researches II, 15.

21 Wyatt L., Approaching Chaos, O Books 2009

the poles; as endorsed by Albert Einstein. This is because difference in gravitational pull of the sun on the ice on the poles, due to the 23.5o list of the Earth on its axis, would be strongest at the solstice. A floating crust, during a period of global warming, would most likely to roll over the poles at the solstice. The greatest likelihood would be about December 21st as that is when the ice on the Antarctic would experience the strongest gravitational pull from the sun. Hapgood pointed out that ice accumulating on a land mass was more prone to cause a crust shift than ice floating on the sea. The prediction of the Maya appears to fit the science.

Historical records from every continent report that the world has fallen over the poles at least four times in the memory of mankind. A major part of stone inscriptions found in the Yucatan refer to this type of world catastrophe. The most ancient of these katun calendar stones of Yucatan refer to great catastrophes at repeated intervals convulsing the American continent. The indigenous nations of the Americas have a preserved memory of these ancient historical events.[22] In the chronicles of the Mexican kingdom it was written: The ancients knew that before the present sky and earth were formed, man was already created and life had manifested itself four times.[23]

The sacred Hindu books, the Ezour Vedam and the Bhaga Vedam share the scheme of expired ages called Yugas, the fourth being the present. They differ only in the time ascribed to each age.[24] The Buddhist Visuddhi-Magga describes seven ages, terminated by world catastrophes.[25]

A tradition of successive creations and catastrophes is found in Hawaii[26] On the islands of Polynesia there were nine ages recorded and in each age a different sky was above the Earth.[27] Icelanders believed that nine worlds went down in a succession of ages, a tradition contained in the Edda[28]

There are seven ages in the rabbinical tradition of Creation: Already before the birth of our earth, worlds had been shaped and brought into existence, only to be destroyed in time. This Earth too was not created in the beginning to satisfy the divine plan. It underwent reshaping, six consecutive remouldings.

22 Brasseur de Bourbourg C., S'il existe des Sources de l'histoire primitive du Mexique dans les monuments égyptiens,

23 Brasseur de Bourbourg C., Histoire des nations civilisées du Mexique I, 53, 1857-1859

24 Volney C., New Researches on Ancient History 1856

25 Warren H., Buddhism in Translations 1896

26 Dixon R., Oceanic Mythology, 1916

27 Williamson R., Religious Beliefs of Polynesia 1933

28 Völuspa, The Poetic Edda, (transl. H. Bellows 1923)

New conditions were created after each of the catastrophes...we belong to the seventh age.[29]

According to the rabbinical authority Rashi, Hebrew tradition knew of periodic collapses of the firmament, one of which occurred in the days of the Deluge.[30] The Jewish philosopher Philo wrote: Great catastrophes changed the face of the Earth. Some perished by deluge, others by conflagration.[31]

In Isaiah 24:1 is written: Behold the Lord maketh the earth empty, and maketh it waste, and turneth it upside down, and scattereth abroad the inhabitants thereof.

If a Samvartaka river of fire, erupting from the sun, were to impact the Earth, as predicted by William Henry,[32] it could trigger a slip of the crust by melting or displacing the ice suddenly on Greenland or Antarctica. Then as described by Charles Hapgood in the Earths Shifting Crust,[33] the crust of the Earth would slide over the mantle and accelerate in the roll over the poles to cause the events described by the ancients without the entire globe going for a tumble as imagined by Velikovsky. In the Hapgood hypothesis the poles don't change. It is the crust turning over the poles that causes the apparent reversal which would cause the change in direction of the sun in the sky.

If the pole positions on the crust reversed, by its rolling over the actual poles of the turning globe, the east-west orientation on the crust would also reverse. A displacement of the Earth's crust over the poles of the planet, could account for the claims in the ancient records that the sun periodically reverses its direction in the sky. This could explain why the South Pole was situated under Hudson Bay less than ten thousand years ago and when the rivers were flowing on the continent, now known as Antarctica, it was in the Northern hemisphere.

A displacement of the Earth's crust over the poles would cause shock waves that would lead to extreme earthquakes and tsunamis throughout the world. Also heat produced by friction as the crust rolls over could cause conflagrations and violent storms described by the ancients. Heat causing evaporation of the oceans followed by condensation when the Earth settles and cools

29 Ginzberg L., Legends of the Jews, 1925

30 Rashi, Commentry to Genesis 11:1

31 Philo, Moses II, x, 53

32 https://www.williamhenry.net/2018/04/ascension-fire-the-solar-flash-and-the-return-of-christ/

33 Hapgood, C., Earth's Shifting Crust: A Key to Some Basic Problems of Earth Science, Pantheon Books, 1958

down would explain the deluges recorded in so many ancient texts. All of this could account for the worldwide catastrophes reported by the ancients, of earthquake and storm, fire and flood that accompany the reversal of direction of the sun in the sky.

We are in a phase of global warming so it is likely we are approaching another crust shift and consequently a pole reversal could occur. In 2014 the European Space agency predicted an imminent magnetic pole reversal but by imminent they meant any time in a few thousand years. However, in Scientific American an article suggested a reversal of the poles could happen sooner than expected.[34]

The hypothesis proposed by Charles Hapgood and endorsed by Albert Einstein in Earth's Shifting Crust [33] links a number of seemingly disconnected events that return to destroy human civilisations and reduce populations. My gravity geyser model is an important piece in the puzzle as it explains the periodic rise of such enormous heat in the Earth that could cause repeating global catastrophes accompanied by the reversals of the direction of the sun in the sky, recorded in the historical annals. The energy of annihilation rising periodically from the centre of the Earth would fit the Hindu tradition of Shiva the destroyer awakening at the end of every yuga.

There is no one better to tell you about the earth changes and ascension than the master of ascension; Yesua. What follows is a transmission[35] from him which I present as a message from a caring friend.

---

34 www.scientificamerican.com/article/earth-s-magnetic-field-flip-could-happen-sooner-than-expected

35 Transmission from the Amethyst Group, Dawlish, UK, 1994

CHAPTER 5
THE ASCENSION PLAN

Greetings my friends. I bring you a message which is important to all of humankind. My heart is full of love for each and every one of you. This is why I am here. I am the Good Shepherd. Some of you will know me in this modern time as Sananda and others will know me as the one called Jesus Christ. In my own day, to my own people, I was known as Yesua.

The time for this planet and the people who inhabit her is running out. Do not be afraid because I bring you a message of hope, not doom and gloom. This planet was created in beauty and loveliness beyond compare and all the people and animals that dwelt therein lived in peace and harmony. Now this wonderful Earth, so beloved by the Father-Mother God, has fallen into great decline and mankind will not listen. Very soon my friends, this planet will not be as you know her. The Earth will change as the All Loving One brings her back to the plan made in the first place.

The world as you know it now – the climate, the famine, the disease, the pain, the torture, the hunger, the greed, the avarice – is coming to an end, and I have come through this instrument to tell you of a plan we have created for those of you who will listen and heed. We have called it The Ascension Plan. The Ascension Plan is for you and for all the peoples of this Earth if they will but heed my word. The Godhead, the Father, so loves this world that he sent me before to show you what could be accomplished if all men would live in peace and harmony. Yet those at that time sought to listen to only parts of my message and other parts were missed and have now been long forgotten, but this planet has still remained in the heart of the Father, his most favoured, and her people on it, his most loved ones.

The Ascension Plan holds your salvation. It holds for you a place of safety. You need only to look at the newspapers, and the reports of fires, floods and earthquakes to realise there is a great change coming upon the planet. Whilst loss of life cannot be avoided at this time, the Ascension Plan shows you how you can overcome this. Throughout the world there are thousands like yourselves who have come to hear the news of the ascension. To many the news is greeted with something akin to remembrance. That they have heard of this long before, but cannot place when. To others it is greeted with abhorrence and stupidity, mockery or cynicism. So be it, my friends, I do not come to make people change their ways of life – their lifelong beliefs, their faith, their religion – I only come to tell them what is happening and how they may help themselves and others like them.

The Ascension Plan is divided into three main parts called the three waves of ascension. The first wave of ascension is already underway. Those already triggered into awakening by the ascension message will be on the first wave of ascension. They will be, for the most part, star-seeds, that is people who have been programmed for ascension even before they incarnated in this lifetime. The first wave we had anticipated would take somewhere in the region of 144,000 people.

The first wave will be a spiritual ascension and will last from 2-48 hours. Those who have desired to go will be taken for the most part in spirit only: some in their sleep, others in a state of wakefulness and some in a state of meditation. On ascension you will be taken to a place which is prepared for you. You will know its love, you will feel its vibration, and you will know the familiarity of the scenery, the place and the people therein. You need not be afraid, all that will surround you will be love and light and the wonderful Father-Mother God, the universal Divine Consciousness, encompassing you with love. Whilst you are there, there will be time for resting, there will be time for re-understanding, dusting down and generally making you feel a whole lot better. You will have come home.

The first wave is divided into two parts, those who desire to return to the Earth plane and those who don't. Each of you, whichever pathway you desire to take, will have the same knowledge of what will take place upon the Earth plane, and if you do decide to return you will be returning as a different person. You will be connecting with your Higher-self so that you can bring back into this bodily form and your environment in which you live, the spiritual essence and some of the things you are capable of but perhaps, as yet, have remained dormant. Those of you who are healers will find that your healing abilities will have been advanced many times. You will find that the healing you give will be almost instantaneous. Those of you who are willing to channel information on the ascension will find many facets of myself available to you – and it is only as in love that you shall receive.

There will be other things as well that will be enhanced within you. You may notice a difference in your bodily form. Your bodies will cease to age and decay in physiological terms; they will begin to regress to their prime. Young people will continue to grow to their prime.

The star-seeds who return back to their environments and families and mundane world, will be ready to assist those who are asking questions. They will be ready to assist groups and explain to them what is happening – to show them, prepare them, ease the way for them and calm their fears so when the second wave begins, they will be ready

So to the second wave. I cannot give you dates as it is not permitted and I would not want to evoke mass panic on planet Earth (it will be in a time of years not centuries) but the second wave is the beginning of the evacuation. The changes that will occur on this planet will be so great and so cataclysmic that it would not be right for any of you to be on her surface when this occurs. This is what the Ascension Plan is designed for. It is a plan of awareness, heightened sensitivity and temporary evacuation. After the first wave, many will have returned to spread the news of the Ascension Plan. We are hoping that this will make things easier for people to leave the Earth for a while because the second and third waves of ascension must be the ascension of the body and the spirit. Those of you who choose to do that; you will afford great assistance to the many who remain here before the final evacuation.

The final wave will be the last chance for ascension. Those who choose to remain behind will miss their chance and will go through the death process. Some will be taken to another place where their souls will be restructured. Some may even return to the dimensions or the other worlds from which they have come, because there are other worlds – you are not the only ones in this vast Universe. The Universes that even you do not know of are innumerable. I could not begin to count them, and within these Universes are planets with beings of light, beings of physical shape much like your own, and beings who just exist in thought and light alone, but beings they are. So some of those will return to the dimensions from whence they came and others will be helped along their way and their pain made bearable, but any at that final wave who express the desire to be with the Father-Mother God then, so be it, we will take them. No one is barred.

The final days will come as soon as the evacuation of the third wave has been completed. The Earth will shift on her axis. She will change – North may become South and South become North and what was wet may become dry and what was dry may become wet. You may see the sun rise in the West and set in the East. There will be a change in the climate and a shuddering and breaking of the Earth's surface. The Earth will be reshaped to my Father's original design. This is truth. This is no story. It is happening now. There will be more flooding, there will be more shaking of the Earth, there will be more fire, and also the destruction of wealth. The trappings which many people think they have here will be destroyed and in their place they will learn these things are not important. As I have said, it is not doom and gloom and it is not going to happen next week. Watch and look around to see what is happening. Understand and with that understanding in your hearts tell those who are still ignorant of what is happening. When the final wave has been completed you may do as you please, you may stay in the created environment or you may return to the new Earth – you may walk between the two.

Your bodily form will be changed. Spiritually you will be changed undoubtedly. You will have no need of vehicles which run on your roads, and again turn to your western Bible, the lion will lay down with the lamb. You are on the brink of a new dawn for all mankind. Some call it the 'New Age'. Some call it the 'Age of Aquarius'. We call it the birth of the world. It is an exciting and wonderful age in which you live and each and You will walk upon the face of the Earth again. Do not think for one moment that in the third and final phase, that is it. Nay, you will return. You will return to see your grandchildren grow but in such a world that only your wildest dreams and imaginings have seen. This in no fantasy, this is no story, this is happening now – your world is changing. With the love that is in my heart and with the love that I bring from the Father-Mother God, will you not heed my words? Will you not look, will you not listen? Tell your friends and help them to understand! Man has progressed; perhaps too much progression. Man has been given the technology and yet he uses it to destroy his world rather than assist it. Man has abused the greatest gift of all which is the world in which he lives. Technology is always being updated and heightened and yet he fails to eradicate the disease which is responsible for killing more human beings than anything else on this Earth plane and that disease is greed. That is why half this planet starves and the other half is at war – you have no idea how much this grieves me. This is why I am here; this is why I have chosen to speak to all who will hear me again. This is not a new message, neither is it a new Sermon on the Mount – it is a plan for your continued race, without it you will surely perish. My Father, the Godhead, will see that no hair on your heads is harmed; this is why He offers to you so freely the Ascension Plan.

My parting words to you are only to think on what I have said. Think them over, think them through and question within yourselves. Try and find others who are of like mind to you and share this with them. Watch the changing face of the Earth and know that I speak the truth. The Ascension Plan is for all of you. Be happy in this and know that if you have the simplicity of love in your hearts, the desire to help your fellow men and women and above all to be part of the change in this world, each and every one of you should partake in the ascension, none shall be barred. Hear my friends, none shall be barred: no race, no colour, no creed, no religion, no Jew, no Gentile. Division shall pass away and in its place will be a union of all humankind and, with that union of humankind, will come the union of life itself, and with the union of life itself the union of the All Loving One and humankind is inevitable. This is what the ascension can do for you. The ascension is given freely and I ask only that you accept it in love and you take it freely. Therefore my loved ones as the Father and the Mother freely give to you, freely receive, freely partake and freely share. Unify your causes; work together in peace and harmony. Let none think he or she is better than another for in the sight of the All Loving One, all are equal. Now I leave you my love and the love of your Father-Mother God who watches over you and waits. May love and light remain always in your lives and I hope, until we meet again, that love will be your goal. Farewell my friends.

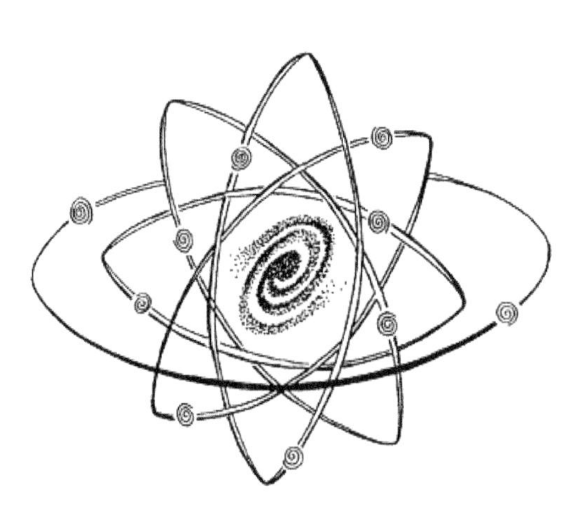

BOOK III
THE QUANTUM VORTEX

"If we do discover a complete theory, it should in time be understandable in broad principle by everyone, not just a few scientists. Then we should all, philosophers, scientists, and just ordinary people be able to take part in the discussion of the question of why it is that we and the universe exist. If we find the answer to that, it would be the ultimate triumph of human reason - for then we would know the mind of God."

- Stephen Hawking
A Brief History of Time
Bantam Press 1988

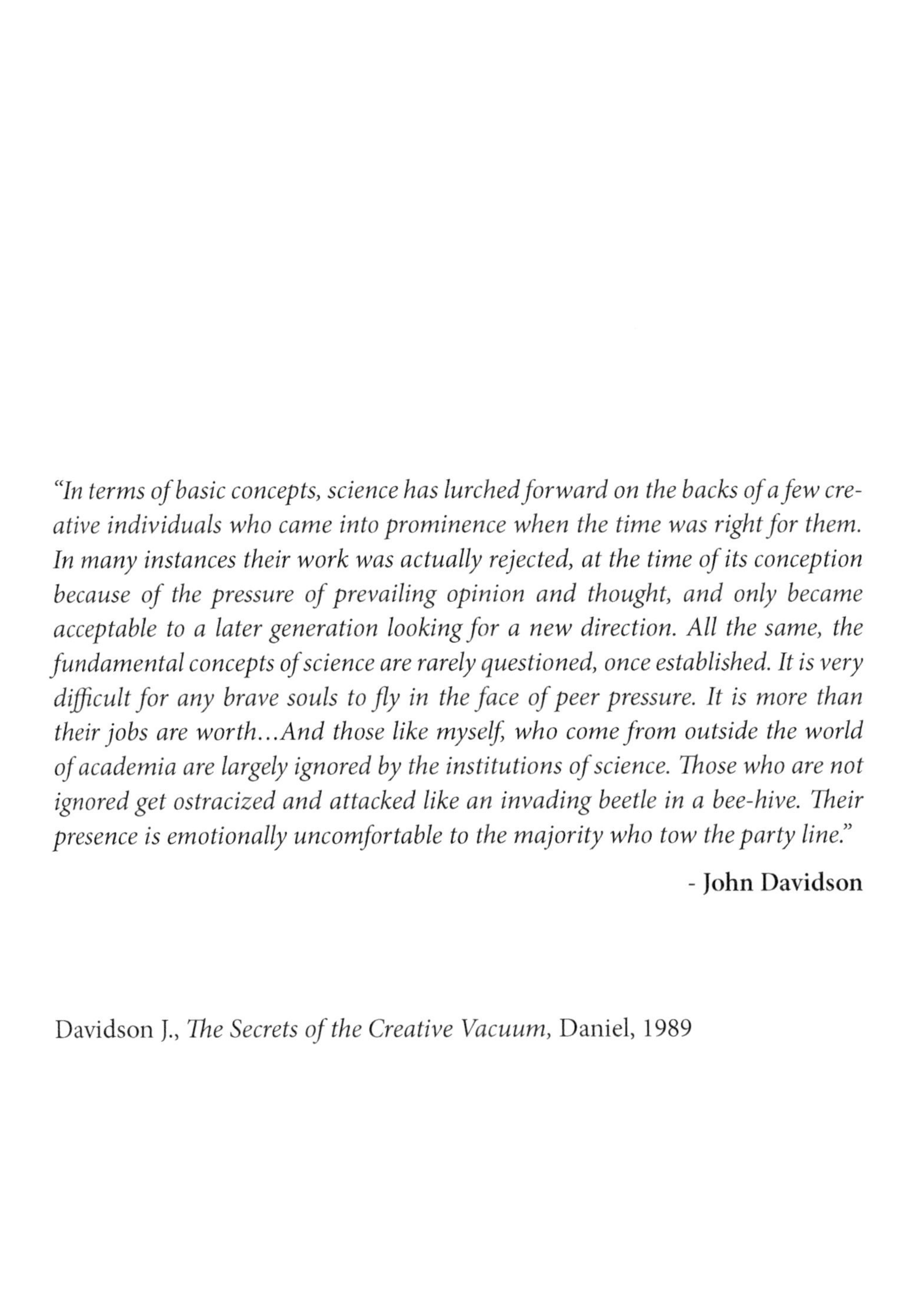

"In terms of basic concepts, science has lurched forward on the backs of a few creative individuals who came into prominence when the time was right for them. In many instances their work was actually rejected, at the time of its conception because of the pressure of prevailing opinion and thought, and only became acceptable to a later generation looking for a new direction. All the same, the fundamental concepts of science are rarely questioned, once established. It is very difficult for any brave souls to fly in the face of peer pressure. It is more than their jobs are worth...And those like myself, who come from outside the world of academia are largely ignored by the institutions of science. Those who are not ignored get ostracized and attacked like an invading beetle in a bee-hive. Their presence is emotionally uncomfortable to the majority who tow the party line."

- John Davidson

Davidson J., *The Secrets of the Creative Vacuum*, Daniel, 1989

INTRODUCTION

In the latter half of the 19th century the popular idea that atoms were solid material particles had been replaced by a theory that they were vortices. The vortex theory for the atom, which dominated physics in Britain at that time, had been used successfully to explain the fundamental properties of matter in terms of vortex motion.

However, early in the 20th century, due to its inability to explain new discoveries about the atom, the vortex theory was abandoned. At the same time the discovery of two smaller particles in the atom, the electron and the proton, revealed that the atom was not the fundamental particle of matter it was thought to be. But no one thought to apply the vortex theory to the subatomic particles. Instead, a team was pulled together in Denmark, by the Danish physicist, Neils Bohr, who had applied quantum theory to the atom and successfully explained the things about the atom the vortex theory had failed to explain. His team seized the initiative and the epicentre of physics moved from Cambridge to Copenhagen.

From 1930 onwards a system called quantum mechanics was developed by the Copenhagen team, which took the world of physics by storm. Back in Cambridge, in 1935, a third particle, the neutron, was discovered in the atom, which provided evidence that the founding principle of quantum mechanics, the uncertainty principle proposed by Werner Heisenberg, was wrong. To save the principle physicists insisted the neutron wasn't what the experiments showed it was. This was a coverup to protect quantum mechanics as almost everyone in physics had jumped onboard the Copenhagen bandwagon. Einstein didn't. He knew quantum theory had taken a wrong turn since the reins had been seized from him by Bohr and Heisenberg but few physicists listened to him. Instead the career physicists went careering off into a quantum quagmire which few could comprehend.

The Copenhagen team drove a final nail in the vortex coffin by discrediting most of the models in physics from the 19th century as naïve realism. I was a naïve schoolboy when I first read about the vortex theory. I realised almost immediately it had been applied at the wrong level. If the smallest particles in matter were vortices, they had to be subatomic particles not atoms.

This book records my endeavour to resurrect the vortex theory at a subatomic level and reveal major deceptions in science. The journey has been truly amazing. Mysteries of the Universe unfolded before my eyes revealing their secrets at every turn. The idea of subatomic particles being vortices of energy explained practically everything in physics, from quantum reality to advanced cosmology, with extraordinary ease and simplicity. Lord Rutherford, who first broke into the atom, said *"These fundamental thing have got to be simple."* That was the keynote throughout my work.

CHAPTER 1
THE VORTEX THEORY

In the 19th Century the German physician and physicist, Herman von Helmholtz, coined the term *quantum.* He also proposed that the smallest particles of matter were vortices in the ether. His theory of the atom as a vortex was championed by Lord Kelvin (William Thomson) who was a leading figure in physics toward the end of the 19th Century.

In Kelvin's day atoms were taken to be the smallest bits of matter which were commonly assumed to be solid material particles, conceived as minute billiard balls. Kelvin despised this idea. To him this material model was unsatisfactory as it offered no explanation for the fundamental properties of matter. He considered the popular, materialistic view of matter to be superficial and naïve. He dismissed the billiard ball model of the atom as totally unsatisfactory: *...the monstrous assumption of infinitely strong and infinitely rigid pieces of matter...Lucretius' atom does not explain any of the properties of matter...*[1] With the endorsement of Lord Kelvin, the Helmholtz theory of the vortex atom replaced the solid particle theory for atoms in the latter half of the 19th Century and was taught at Cambridge until 1910.

James Clerk Maxwell, who developed the equations for electromagnetic theory, was a strong proponent of the vortex atom. In the Encyclopedia Britannica of 1875 he wrote: *...the vortex ring of Helmholtz, imagined as the true form of the atom by Thomson (Lord Kelvin), satisfies more of the conditions than any atom hitherto imagined...*

J. J. Thomson (1856-1940), who discovered the electron, was professor of physics at Cambridge when he said: "*...the vortex theory for matter is of a much more fundamental character than the ordinary solid particle theory.*"[2]

With the vortex theory physicists were able to reduce the properties of matter

1 Thomson W. Popular Lectures and Addresses 1841

2 Thomson J.J. Treatise on the Motion of Vortex Rings, Cambridge University, 1884

to the dynamic principle of vortex motion. But the Victorian vortex theory failed because it was applied to the atom and not to subatomic particles. That was not the fault of the 19th century scientists because in their day the atom was thought to be the smallest particle of matter. Subatomic particles were not yet discovered.

In 1964 I read about the ether theory of the vortex atom in *An Advanced Course in Yogi Philosophy and Oriental Occultism*[3] by William Atkinson, penning as Yogi Ramacharaka. Initially I wasn't impressed because the ether had already been disproved when the book was first published in January 1905. But when I read that yogis had perceived the smallest particles of matter to be vortices of energy I was realised I had discovered something of immense importance. I knew that Albert Einstein had declared the equivalence of mass and energy later in that same year. If yogis had anticipated Einstein, then maybe they were correct in their description of fundamental particles of matter as vortices of energy.

Kelvin used smoke rings as his model for the vortex atom. He presented it as a toroid vortex but I realised this model would not work for corpuscular sub-atomic particles. The quantum vortex of energy had to be a spherical vortex. This was obvious to me because of the degrees of freedom for energy spinning freely on a single point. Energy would naturally form a ball vortex rather than a toroidal vortex. I chose the ball of wool as my model for subatomic vortices of energy as I realised it was the freedom of wool to wind in every direction, on ever changing axes, that set up a wool-ball vortex.

I discovered the wool-ball model in December 1968. I was twenty at the time and in my first year at Queens University of Belfast. During the Xmas vacation I had been billeted with a friend of my father's, in a caravan at Crackington Haven in Cornwall. He asked me to hold up a hank of wool so he could wind it into a ball. I was mesmerized as I watched him. I saw a ball vortex forming in his hands as he spun the wool on constantly changing axes. I realised the wool-ball was a vortex when it was being wound or unwound. In my first term at Queens my vortex theory had been challenged in the physics department on the grounds that vortices have an axis of spin and subatomic particles have no poles that would reveal any axis of spin. As I watched the formation of the ball of wool I realised the subatomic particle could be a ball vortex without measurable poles. The axis of spin of a normal vortex would not be apparent in a ball vortex of energy as the spin would be on constantly changing axes.

---

3 Yogi Ramacharaka, An Advanced Course in Yogi Philosophy and Oriental Occultism, The Yogi Publication Society, 1904

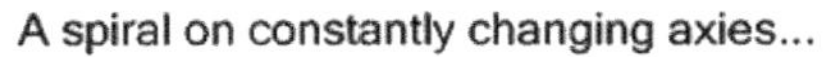

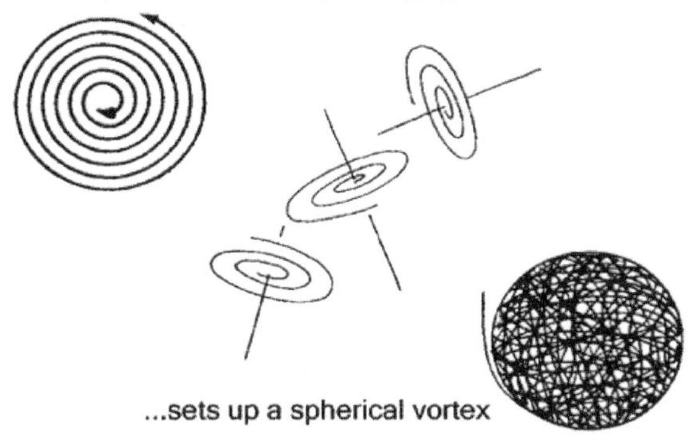

The wool ball model revealed how vast amounts of energy could be stored in minute amounts of matter. A ball of wool is a compact form of wool. Likewise, a spherical quantum vortex would be a very compact form of energy. If a ball of wool is unraveled it releases a lot of wool so if nuclear vortices were unraveled, they would release vast amounts of energy. This explained the nuclear explosion. By defining mass as quantity of vortex energy I could explain the greatest enigma in physics; the equivalence of mass and energy.

Appreciating how the subatomic vortex could store energy as mass I could understand $E=mc^2$. I saw how vast amounts of energy could be stored in the mass of minute subatomic particles of matter. In time, with the vortex model, I would also come to understand how *potential energy* could be stored in the nuclear binding force and the forces of gravity, magnetism and electric charge. I was able to explain the cause of these forces as interactions between dynamic vortices of energy.

The quantum vortex theory for subatomic particles could be classified as a string theory. Most string theories attempt to account for everything in terms of *strings of vibration.* The quantum vortex contribution to string theory is to treat subatomic particles as *strings of spin.* A vortex of energy would not be a spinning string. It would be just spin. Strings are just a model to help us imagine particles of energy as *lines of the movement of light* either vibrating as a photon of light or spinning to form a subatomic particle.

The vortex revealed how non-substantial energy can set up an illusion of apparent materiality. If material substance is defined as inertia, mass, potential energy and three-dimensional extension then because these are properties of the quantum vortex of energy, what people experience as material substance can be explained away as spin. People who believe only in what they can see and touch are deluded. At a quantum level there is nothing there. What they think is solid and substantial is nothing but spin.

The quantum vortex is a three-dimensional (3D) spiral. The 3D spin of energy, forming subatomic particles confers three dimensionality on matter. This explains why the world we live in is three dimensional. That is something that most of us take for granted.

The *static inertia* of matter is its tendency to stay put unless forced to move. This is conferred on matter not by material substance but by the spin of energy forming the quantum vortex.

Spin sets up inertia. This is illustrated by the gyroscope. The spin of a gyroscope resists motion out of the plane of spin. This is also depicted by the spin of a pebble on a pond. The spin of the pebble prevents the pebble falling out of the plane of spin and into the water, which enables it to skip on the surface of the pond.

Richard Feynman said: *"The laws of inertia have no known origin."*[4] The quantum vortex provides a simple and straightforward account for inertia.

The spin of energy on infinite planes, forming a spherical subatomic particle of matter, explains the inertia of mass; why matter resists movement in any direction. Static inertia is conferred on the atoms, crystals and molecules in bodies of matter by the static inertia of the quantum vortices of energy forming them and also by their spin in atoms.

If the quantum vortex of energy is a *system of static inertia* the quantum of wave-propagating energy would be a system of kinetic inertia. The laws of inertia can be summarised in a single statement. The quantum vortex of energy stays put unless it is moved whereas the wave train quantum of energy keeps moving unless it is stopped. The reason for this is particles of energy are particles of activity and the innate tendency of energy is to maintain its state of motion. Motion is fundamental to everything because the motion called energy, not material substance, is the essence of everything.

In essence the theory of the vortex atom was that vortex motion is fundamental to the smallest particles of matter. It was abandoned at the turn of the twentieth century because it was unable to provide an account for the line spectra of atoms. The Danish physicist, Niels Bohr, explained spectral lines by applying quantum theory to the atom. He suggested the atom was a system of electrons, in distinct orbits, that could move into higher orbits. Spectral lines, Bohr explained, represented the energy signature of an atom as its electrons absorbed energy to make quantum leaps from lower to higher orbits and emitted energy as they fell back to their original orbits.

---

4 Feynman R. The Character of Physical Law, Penguin 1992

With the demise of the theory of the vortex atom, the centre of physics moved from Cambridge to Copenhagen. There a young German physicist, Werner Heisenberg, took a quantum leap in his thinking. In the early 1920s he suggested it was only possible to be certain that electrons exist, in a quantum leap, when they absorb or emit light. This insight came to him when he watched a man walking in a park after dark. The man was apparent only when he was in the light of the gas lamps but disappeared when he was striding between them.

Heisenberg's brain-wave headed the Copenhagen interpretation of quantum reality that particles can only be assumed to exist when they are observed. The idea that quantum reality depends on conscious observation is fundamental to quantum physics. It led the father of quantum theory, Max Planck, to remark, *"I regard consciousness as fundamental. I regard matter as a derivative of consciousness."* Increasingly quantum physicists were thinking that consciousness rather than material substance might be the bedrock of reality. Then two decades later the non-substantial nature of quantum reality was firmly established in cosmic ray research.

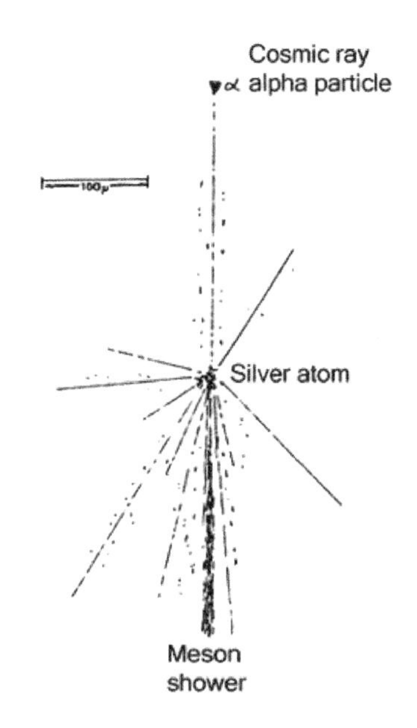

Cecil Powell, professor of physics at Bristol University developed specialised photographic plates for the study of cosmic rays. In 1946, one of his plates caught a high energy cosmic ray particle slamming into the nucleus of a silver atom in the photographic emulsion.[5] When the cosmic particle was stopped by the nucleus its kinetic energy continued on to be transformed into a shower of 140 neutral particles of matter called *pi-mesons.*

Powell's cosmic ray photograph witnessed the transformation of kinetic energy into mass. It provided experimental evidence that the Universe is made of energy, which is pure movement. Energy has no substance. There is no underlying material substance in energy that moves. This discovery overturned the material philosophy of Democritus. Subsequent high energy research has piled on the evidence that the philosophy of materialism is unsound. It is spin!

5 McKenzie A. E. A Second MKS Course in Electricity, Cambridge University Press, plate 19, 1968

Materialism is not supported by experimental science. The idea of the atom, which survived for the benefit of science in a Latin poem by Lucretius, was brilliant but Democritus was deluded in his assumption that it was something substantial. We can forgive him because in the vortex theory we can now see how easy it is to be deceived by the 3D spin of energy into thinking the smallest particles of matter are substantial things. They are nothing but energy which is unsubstantial. Energy is activity. Think about it. After the crash of the cosmic ray particle into the silver atom it was no longer moving. Its progress was arrested by the nucleus of the atom as though it had hit a brick wall. But its kinetic energy went on. This activity of the cosmic ray particle left the arrested particle behind and plunged on through the nucleus. Then the amazing happened. The animation of the cosmic ray particle exploded out of the nucleus, not as heat or light but as a litter of new particles of matter. Energy as pure animation had been transformed into mass.

In his speculative materialism Democritus taught material atoms are fundamental to matter and scientists in the modern era defined energy as a measure of their activity. His atomic hypothesis could never accommodate the discoveries of modern physics that atoms can be split into subatomic particles and that these are formed of energy. It takes the vortex theory to explain how the smallest particles of matter could be energy in the form of spin and show how spin in the vortex can explain the fundamental properties of matter such as inertia and three dimensional extension.

Until the advent of high energy nuclear physics, materialism had the day. Scientists believed that energy was a measure of the activity of things and energy could not exist without something that is active. It appears to be obvious that without a body there is no movement. If you are leaping around and then stop the leaping stops too. The leaping doesn't go on without you but in quantum physics it does.

Cosmic ray research and subsequent high energy physics has established that the movement of high energy particles can exist without the particles. In Professor Powell's experiment, after the cosmic ray particle smashed into the nucleus of the silver atom it was arrested but its movement went on through the nucleus without it and came out the other side as a shower of new particles of matter. The cosmic ray experiment established, at a quantum level, the psyche principle of Plato. The animation of a body can exist after the demise of the body and go on to form new bodies.

Democritus, said *"Only atoms moving in space exist, everything else is opinion."* Modern physics has revealed that is just an opinion. The equation $E=mc^2$ has revealed that the same energy which moves everything about is ultimately what everything is made of. Energy is neither created nor destroyed and energy transforms endlessly from one form to another. The models of the quantum wave and the quantum vortex for particles of energy can help us appreciate the transformations of energy into mass depicted in cosmic ray and high energy research.

People have been told that mass is a form of energy but while physics has provided equations and the evidence for this, up until now there has been no simple visual model in physics to match the naïve realism of solid material atoms taught by Democritus and promulgated by Lucretius. The idea of material particles cannot explain how mass is formed of energy and how mass can be transformed back into energy whereas the quantum vortex can show how energy forms mass and how it can be released from mass and how spin can set up the illusion of material substance underlying mass.

CHAPTER 2
MASS

In physics today Higgs boson is thought to be responsible for mass. The Higgs boson theory for mass is incredibly complex. It requires advanced math and a substantial understanding of quantum mechanics to grasp it. The description of mass in the theory of the quantum vortex is much simpler and easier to understand. It can be summarised in four words: Mass is vortex energy. As Lord Rutherford, the father of nuclear physics said, *"These fundamental things have got to be simple."*

To appreciate the three-dimensional quantum vortex of energy as mass we need to expand Einstein's equation $E=mc^2$. E quantifies energy, the symbol c denotes the speed of light and m represents mass.

Imagine you go into a fruit shop and ask for three. The shopkeeper could not serve you because he wouldn't know whether you wanted three apples or three oranges. If you went in and asked for apples, he couldn't serve you until you said how many apples you wanted. If you want three apples he could only serve you if you ask for three apples. To define something properly it is necessary to stipulate what it is and how much of it there is.

In Einstein's equation, E tells us how much energy we have whereas c tells us what energy is. Energy is the speed of light. This is not an easy concept to grasp because energy is not a thing. It has no mass and no substance. In essence a particle of energy is a particle of speed with a form which is either vortex or wave. The speed of light is measured as 299,792,458 metres per second. But the speed of light c is not an ordinary speed. It is a constant. Everything in our world is relative to the speed of light. This is because it is the fabric of our world. The speed of light spinning is the basis of matter and vibrating it is the basis of light. The speed and the form of energy – spin or wave - is very stable, which gives energy its definitive quality of being. Energy is no thing but it is not nothing. It is very real for us as our existence depends on it.

In the fruit shop analogy a full description for the apples we want is three apples, likewise a full description of energy would be Ec. If this complete definition for energy is applied to Einstein's equation, then $E=mc^2$ would expand to $Ec=mc^3$. It is tempting to simplify the equation back to $E=mc^2$ but that would not be appropriate because c is an invariable constant necessary for qualifying not quantifying the fundamental nature of physical reality. This expansion of the equation describes mass as c^3, the distribution of energy in three dimensions.

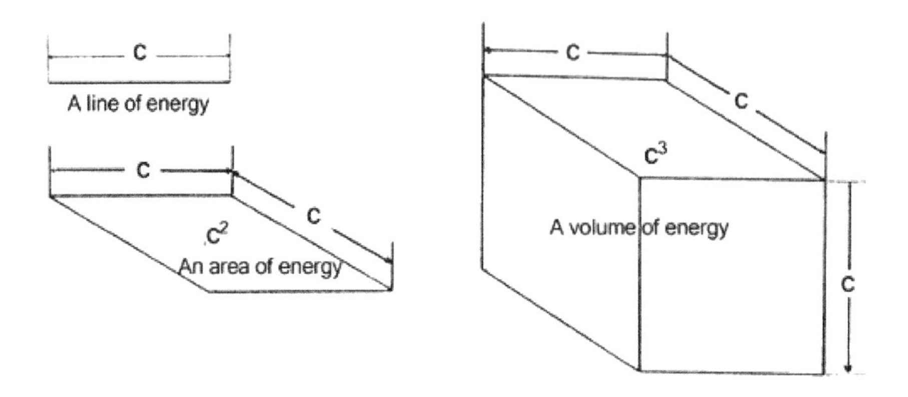

c describes a line of the movement of light in one dimension. c^2 describes the movement of light in two dimensions. c^3 describes the movement of light in three dimensions. The expanded equation $Ec=mc^3$ defines mass as a three-dimensional form of movement of light.

Three-dimensional extension is a fundamental property of matter. Common experience reveals every massive object has three-dimensional (3D) extension. Most people accept this without question but when energy is described as a line of the movement of light in 3D spin, Einstein's equation reveals why mass extends in three dimensions.

In the quantum vortex theory mass is defined as quantity of vortex energy. Mass is spin. Mass as 3D spin is a property of the vortex, not the energy within the vortex. Photons do not possess mass because they are the wave form, not the vortex form of motion.

Momentum is defined as mass times velocity. Momentum laws that apply to a vortex particle in matter do not apply to the energy in the vortex forming that particle. For example, the law of conservation of angular momentum can be applied to the motion of vortices because the vortices have mass but the law cannot be applied to the energy in vortex motion because momentum incorporates mass and energy has no mass.

Mass is associated with *static inertia.* This inertia of mass is a consequence of the 3D spin of energy in the quantum vortex. Exemplified by gyroscopic spin, the spin of energy on infinite planes in the quantum vortex sets up resistance to movement in every direction simultaneously. The inertia of mass is a fundamental property of matter. This defining property of material substance is explained away as spin in the theory of the quantum vortex.

The model of the quantum vortex can also account for potential energy. Potential energy is the energy stored in mass and the forces associated with massive particles. The spin of energy in the quantum vortex enables energy to stand on a single point. Spin is the standing form of energy that enables energy to be stored in mass.

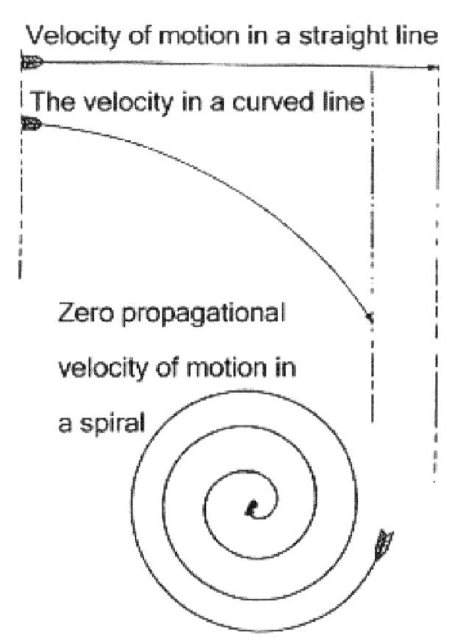

A straight line is a velocity *vector* representing maximum speed in a straight line of movement. If the line begins to follow a curved path the velocity vector in a straight-line would diminish. If the line of motion then spins in a spiral path its velocity in the straight line – its propagation vector - would be zero. This is how the spiral motion of energy in spin could set up a system of apparent stasis, how vortex motion could set up the passive inertia of mass and how energy can be stored as potential energy.

To summarise, in the vortex theory energy exists in a propagating state or standing state depending on whether it is vibrating in waves or spinning in a vortex. Energy has no substance. It gives rise to the illusion of substance through spin. Energy exists as a constant state of motion either in a train of propagating waves that sets up a state of kinetic inertia, or a standing system of 3D spin which sets up mass and its associated state of static inertia.

The models of wave vibration and vortex can help us to better understand energy. Vibration or spin are forms of motion which confer on a quantum of energy either an active or a static state but both states are fundamentally dynamic because activity is the underlying nature of energy.

The wave vibration and vortex models for the quantum also enable us to appreciate inertia as a fundamental property of energy arising from its dynamic state. In these models we see propagating energy as a state of action and standing energy as a state of rest setting up active and passive inertias.

The vortex and wave vibration models help to clarify our understanding of energy and inertia. If we think of inertia only in terms of the static state we may be bewildered because if everything is energy how can anything be static. The vortex model enables us to appreciate how the static state originates from the dynamic state of energy in spin motion and the vibrating wave-train model helps us see how wave-kinetic inertia originates from the dynamic sate of energy in wave-propagational motion.

When Richard Feynman said "*...the laws of inertia have no known origin*", *and "It is important to understand in physics today we have no idea what energy is,"* I realised we can only know the origin of the laws of inertia when we know what energy is. There is elegance and simplicity in the appreciation of mass as the spin of energy in the quantum vortex, especially in the straightforward accounts for electric charge and magnetism as effects of the dynamic spin of energy extending beyond massive particles into infinity.

CHAPTER 3
ELECTRIC CHARGE AND MAGNETISM

In the vortex theory the forces associated with subatomic particles of matter are considered to be vortex interactions acting as an extension of the dynamic mass of the vortex of energy. Quantum vortices of energy are dynamic. When they overlap they interact and these interactions are thought to account for the forces of electric charge, magnetism and gravity.

The vortex account for forces begins with the assumption that energy is neither created nor destroyed which implies there would be no limit to the extension of a vortex of energy. Vortex energy can diminish in its intensity but never vanish altogether. This is illustrated by the principle that one can never get rid of a pie by dividing it into smaller pieces, it just gets more dispersed. As the concentric spheres of vortex energy expand, the intensity of energy in them diminishes but never goes to zero. Like the pie, vortex energy may spread out more thinly but it never ceases to be.

To appreciate vortex interactions imagine you are in a quantum vortex and experience it as a fireball. As you move outwards from the fiery centre the energy rapidly diminishes. Suddenly it vanishes altogether, as though you have come to the end of the vortex and broken out of its fiery domain. However, the apparent surface of your vortex is not its boundary, it is just the last intensity of vortex energy that you can perceive. Looking out from this point into the darkness, you see other fiery vortices moving about. You assume that all these spinning balls of light are separate vortices, occupying a void of darkness - like stars in the night sky. However, your senses deceive you. The apparent void of darkness is full of energy extending from all the vortices. As the invisible energy from one quantum vortex overlaps that of another, there is an interaction between them. Standing on your vortex, seeing nothing but a void between you and the others, you might be perplexed at the inexplicable attractions and repulsions between them and speak of these as *action at a distance*.

As vortices of energy, quantum particles of matter have no actual bounding surface so every vortex is overlapping every other vortex in existence. Vortices of energy are intrinsically dynamic and because of their infinite extension they always overlap and so are in a continual state of interaction. The interactions between vortices of energy, extending beyond our perception, account for the ability of primary particles of matter to act at a distance. It is the infinite extension of subatomic vortices of energy that sets up infinitely extending fields of force.

Richard Feynman said, *"It is simple therefore it is beautiful."* Compare the simplicity of the vortex theory for electric charge with his QED account. Feynman proposed that virtual particles of light pop up between charged particles and cause the interactions between them. After the exchange with another charged particle the virtual photons vanish again. As the whole process of the light appearing and disappearing again occurs within the bounds of Heisenberg's uncertainty principle the particles of light can never be detected as such. This is why they were called *virtual photons.*

Analogous to broadsides between battleships the virtual photons fired between charged particles were considered by Feynman to be responsible for the forces of electric charge and magnetism pushing them apart or pulling them together. Maxwell's electromagnetic theory for light led physicists to believe that the electric and magnetic fields of force associated with matter are linked to light. For physicists like Richard Feynman it made sense to propose that electric and magnetic forces associated with matter were caused by the exchange of photons of light.

Richard Feynman drew diagrams to illustrate the exchange of virtual photons. These Feynman Diagrams have also been used to illustrate the exchange of other force-carrying particles to explain the action of other forces in the theories of quantum mechanics.

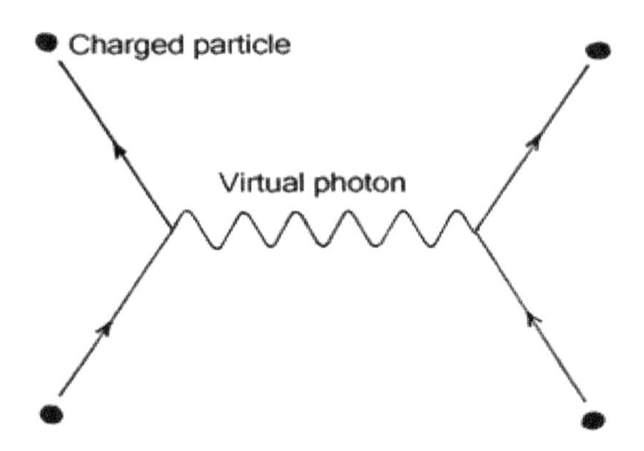

The concept that force-carrying particles could be exchanged over large distances met with certain difficulties. Once a force-carrying particle launched into space how could it locate its target particle at a distance? This was essential if it was to cause an interaction, decay and quickly repay the debt of energy incurred in its formation within the time constraints of the uncertainty principle. Energy conservation laws didn't allow for any force carrying particles to be lost. To overcome this problem *gauge theory* was introduced which suggested that force-carrying particles could gauge the force, that is they could feel for the presence of another force-carrying particle and so home in on their target. There are problems associated with this theory. The idea of particles homing in on each other over short ranges could be imagined but using gauge theory to account for electromagnetic fields operating over the infinite distances of outer space beggars belief. It is hard enough to imagine the instantaneous exchange of virtual photons through deep space let alone their ability to gauge for each other over vast distances. This is an example of where quantum mechanics satisfies the math but defies common sense.

In the quantum vortex theory there are no problems of distance for forces caused by vortex interactions because vortices of energy are infinite extensions. No additional arbitrary elements are necessary in quantum vortex theory to account for the ability of the forces of electric charge and magnetism to act over infinitely large distances if they result from infinite vortex extensions.

The idea that virtual photon exchanges between particles could account for the fields of electric charge posed another problem. Harald Fritzsch, pointed out that if protons possessed energy for electromagnetic exchanges they should be more massive than neutrons. *"We do not understand why the neutron is heavier than the proton. Indeed, an unbiased physicist would have to assume the opposite by the following logic. It is reasonable to think that the difference in mass between the proton and neutron is related to electromagnetic interaction since the proton has an electric field and the neutron does not. If we rob the proton of its charge, we would expect the neutron and the proton to have the same mass. The proton is therefore logically expected to be heavier than the neutron by an amount corresponding to the energy needed to create the electric field around it."*[1] This problem does not occur in the quantum vortex theory where electric charge is an expression of the dynamic inertia of spin in the quantum vortex of energy. Quantum vortices have no requirement for additional energy to account for their charge interactions if these are an expression of their inertia. In the quantum vortex theory a neutron is treated as an electron bound to a proton. If that is true then the mass of the electron would account for the neutron having a greater mass than a proton.

---

1 Fritzsch H, Quarks: The Stuff of Matter, Allen Lane 1983.

As the quantum vortex extends in three dimensions the drop in intensity of vortex energy would obey the *Inverse Square Law*. This applies to all three-dimensional extensions. The inverse square law directs that *the intensity of vortex energy at a distance from the centre of the vortex would be inversely proportional to the square of that distance.*

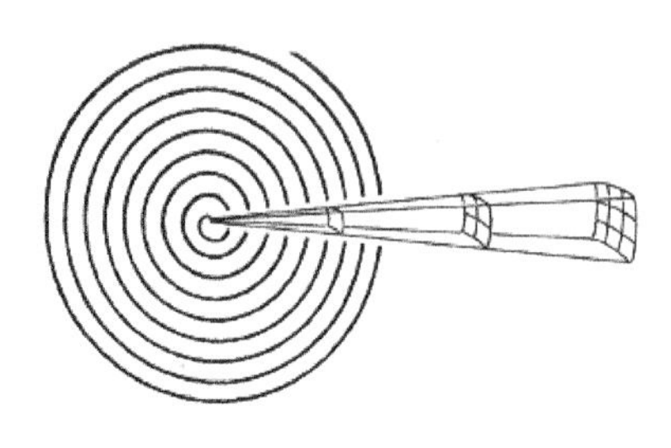

This law indicates that as the distance from the vortex centre doubles the intensity of vortex energy drops to a quarter of its original value. As the distance trebles the intensity drops to a ninth. The law also shows that with distance, while the intensity of energy rapidly diminishes it never falls to zero. The fraction may become small, even infinitesimal, but it never ceases to be. This supports the idea of the infinite extension of the quantum vortex. The concept that every electron occupies the entire universe could provide a simple account for the non-locality of electrons and *Einstein's EPR paradox* - the way subatomic particles affect each other instantaneously over infinite distances.

The Inverse Square law applied to vortices of energy would account for Coulomb's Law. In the eighteenth century the French scientist, Charles Coulomb first observed that the force of interaction between charges is inversely proportional to the square of the distance between them.

In the spherical vortex there are only two ways in which energy can flow - into or out of the centre. This property of the quantum vortex could account for the opposite signs of charge, positive and negative. I arbitrarily represented positive charge as the flow of energy out of the centre of the quantum vortex and negative charge as the flow of energy into the centre.

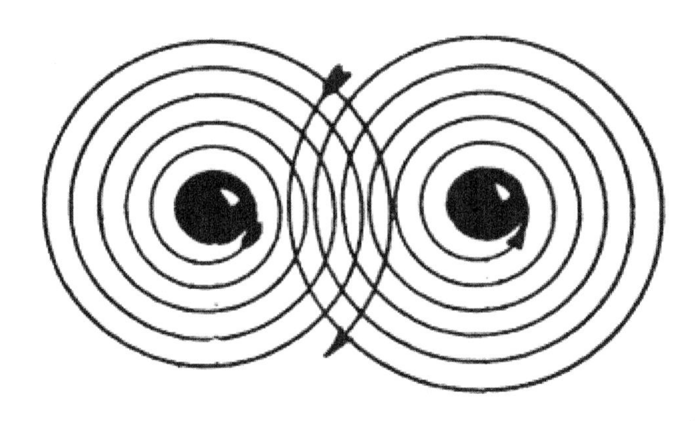

When electric charge is treated as a vortex interaction, each quantum vortex would represent an indivisible quantum unit of interaction. As an irreducible quantum of electric interaction charged particles of matter, as vortices of energy, would all have the same unitary value of electric charge. This is evident in nature where all charged particles have the same value of charge. To appreciate this point, think of an army. An army is made up of individual soldiers. It increases or decreases by reinforcement or loss of whole numbers of men. It doesn't matter whether the soldiers are big or small; each man carries a rifle which enables him to act at a distance. His effectiveness is independent of his mass and just as soldier with a fraction of a rifle would be ineffective so elementary particles have never been observed with a value of charge greater or lesser than unity. There are no fractional charges in nature. Electric charge builds up as an integer multiple of single units of charge. This fits with the quantum vortex theory that a unit of charge is the action of a single vortex of energy, which has nothing to do with its mass.

In nature electric charge accumulates by the addition of individual charged particles regardless of their mass. This fits with the quantum vortex theory. Charges are either positive or negative – representing opposite direction of energy flow in or out of the quantum vortex. Charge is increased or decreased by the accumulation or loss of individual charged particles. Equal numbers of opposite charges cancel out charge. The accumulation, loss, or cancellation of charge always occurs in whole units of charge.

In quark theory, where neutrons are not thought to be electrons bound to protons, there is an arbitrary assumption of fractional charge. Up-quarks are imagined to have ⅔ charge and down-quarks ⅓ charge. The proton is supposed to be formed of two up-quarks and one down quark contributing ⅔ + ⅔ - ⅓ fractional charges which = 1. This is how a proton is imagined to have unitary charge. The neutron is thought to be formed of two down quarks and one up quark each contributing ⅓ + ⅓ - ⅔ fractional charges = 0. In this ludicrous theory that is how a neutron is supposed to have no charge. Physicists have accepted without question that quarks have fractional charges when there is no evidence for fractional charges in nature. Quark theory is clearly crazy.

The quantum vortex fits with nature and its properties are simple, obvious and self-evident. For example, in its distant action as electric charge the spherical quantum vortex behaves like a system of discontinuous concentric spheres whereas closer to its centre it displays its properties more as a continuous 3D spiral. This is evident in a ball of wool. In a ball of wool, the three dimensional spiral of wool is tighter and more pronounced near its centre but as the wool ball increases in size, its continuous spiral nature tends to diminish and increasingly it appears more as a discontinuous system of concentric *nested* spheres. This duality of continuity and discontinuity is evident in the quantum vortex. It is reflected mainly in the distance from the centre in the vortex.

Electric charge is a result of the behaviour of the quantum vortex as it acts at distance from its centre on other vortices. In this distant situation it behaves as if it were discontinuous, expanding or contracting, concentric spheres of energy. In an atom, where the vortex of energy is already vast in its extension, electric interactions occur as the effects of expanding or contracting concentric spheres of energy and magnetic interactions appear to be the effects of rotating concentric spheres of energy. These are the effects of one extending quantum vortex acting at a distance on another.

The spiral nature of the quantum vortex is not evident as an action of the vortex. Close to the centre of the vortex it is evident as an effect its shape as a 3D spiral has on a wave quantum of energy. In the nucleus of an atom, closer to the centre of the quantum vortex, the 3D spiral is very evident in its effect on wave forms of energy. In upcoming chapters I explain in detail what happens to a wave-train quantum of vibrating energy when it encounters a quantum vortex of spinning energy, how this effect can define the circumference of the nucleus and subatomic particles and how it can account for kinetics, wave-particle duality, the strong nuclear force and nuclear energy.

Because electric charges appear to be growing or shrinking concentric spheres of energy, when they act on each other it is self-evident that their interactions would be effective along their radii. It follows, therefore, that electric charge interactions can be described as the radial effects of vortex energy or vortex *radial vectors.* A *vector* is the direction in which a motion is effective.

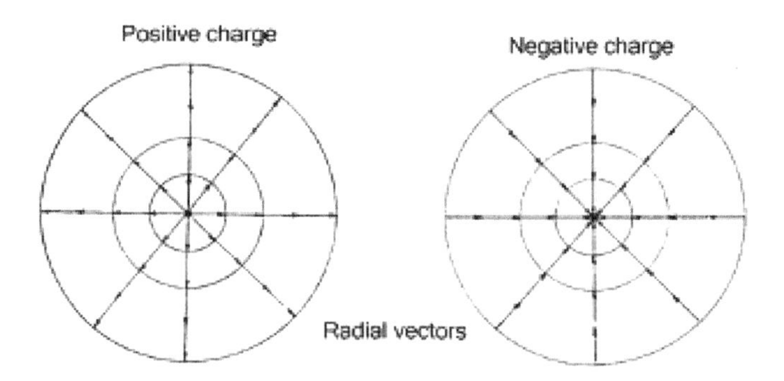

The French scientist, Charles du Fey recognised that like charges repel and unlike charges attract. These characteristic interactions of charges can be accounted for in the quantum vortex theory simply by plotting the radial vectors of overlapping concentric spheres of vortex energy.

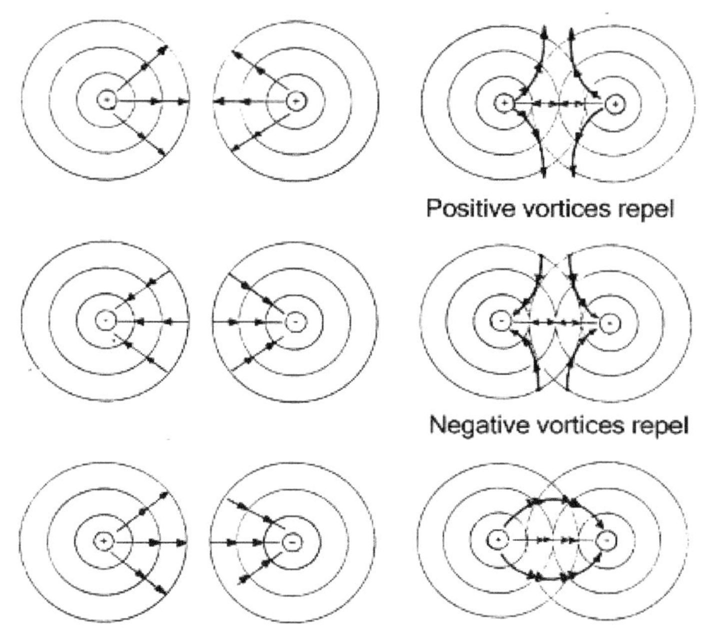

The points of interaction between overlapping radial vectors of vortex energy can be used to plot the lines of force acting between charged particles. These plots correspond to textbook diagrams of the characteristic electric lines of force acting between charged particles.

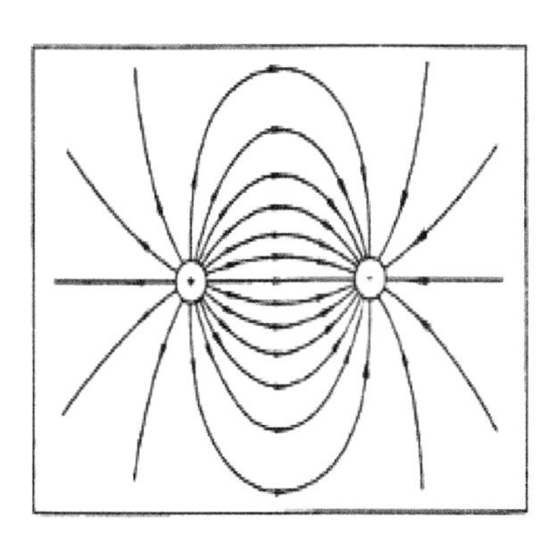

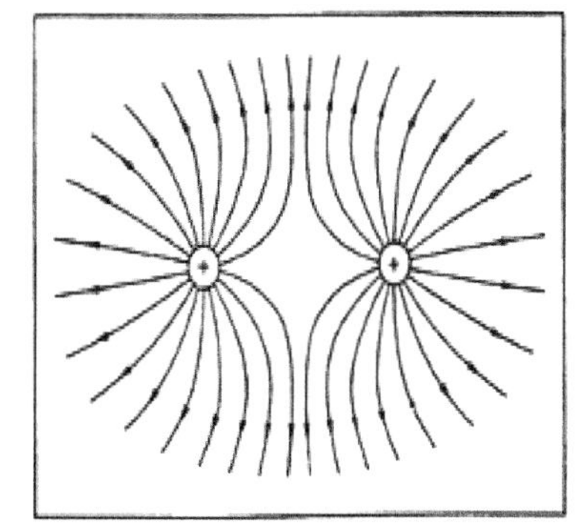

If the flow of energy in two interacting vortices is opposite in direction – if they are both expanding or both contracting - they act *against* each other and repel. If the flow is in the same direction – if one is expanding and the other is contracting - they act *with* each other and attract. This is why particles with the same charge repel and particles with opposite charge attract. Mid-way between attraction and repulsion is zero interaction and mid-way between movement in the same and opposite direction is movement at right angles.

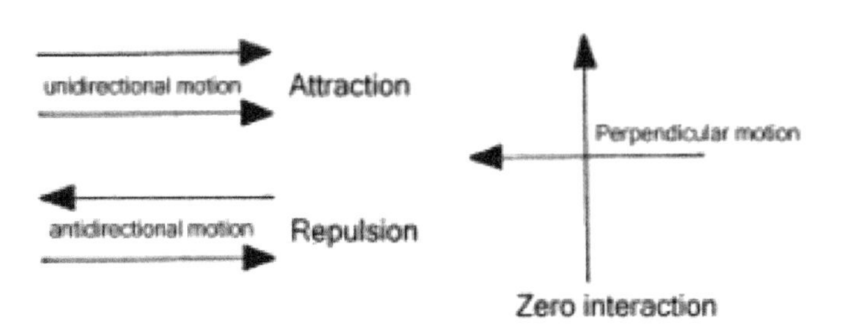

Perpendicular vectors set up the distinction between electric and magnetic fields. Charged particles set up magnetism if they move at right angles to their charge which is why the spin of electrons in atoms gives rise to magnetic moments.

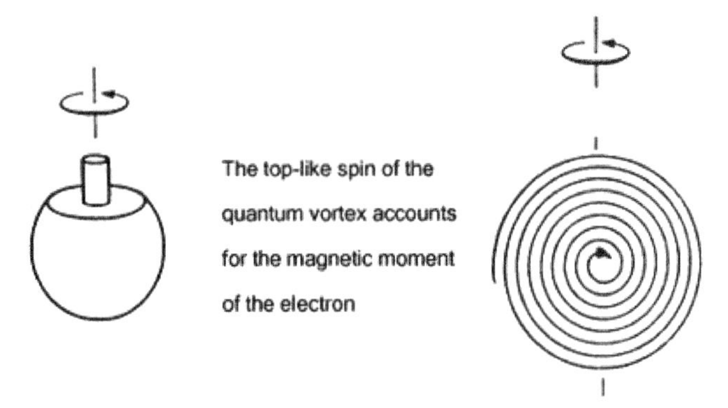

The magnetic moment of an electron results from its rotation. If the electron is a quantum vortex the rotation causing its natural magnetism would be perpendicular to the contraction of the concentric spheres of vortex energy that forms it and its field of electric charge. Two types of spin, one analogous to a whirlpool and the other to a top, would set up the two distinct force fields of charge and magnetism.

If charge is the growing or shrinking of the concentric spheres of vortex energy along their radii, the rotational spin of the spheres, causing magnetism, will be effective at a tangent to them.

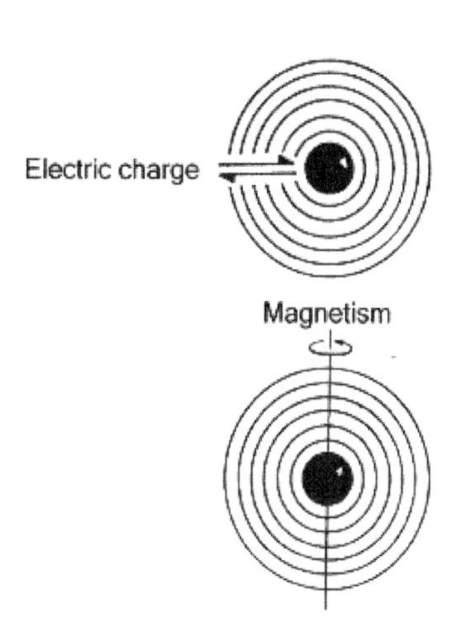

Because a tangent is at right angles to the radius of a sphere, the rotation of vortex energy causing magnetism would be effective at right angles to the flow of vortex energy causing its charge. In the vortex theory wherever vortex vectors act at a tangent to the concentric spheres of charge they set up magnetism.

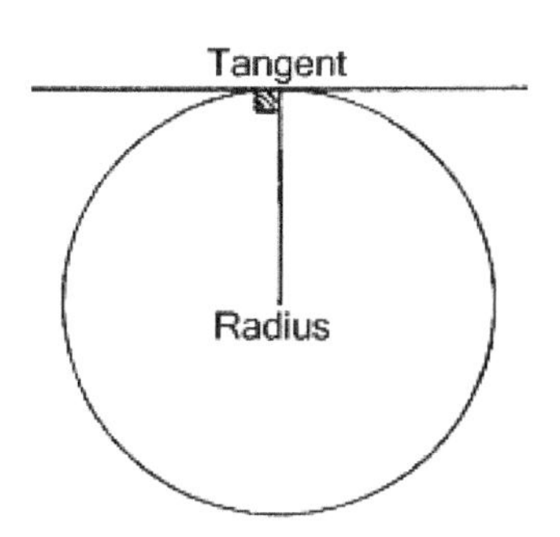

Imagine children in a park blowing balloons while playing on a roundabout. Close your eyes and visualise the scene. The motion of expansion of a balloon would be perpendicular to its motion on the roundabout. So it is the primary *in or out spin* in a vortex of energy forming a subatomic particle would be perpendicular to its secondary *rotational spin.*

If vectors of energy set up force fields and perpendicular vectors of energy create distinct fields and if when the fields are perpendicular they do not interact with each other, we can understand from this why photons are stable. Photons consist of two vectors of energy vibrating at right angles. A lack of interaction between these perpendicular fields of energy could explain how they coexist without interference.

Because the perpendicular vectors of energy in photons interact with the perpendicular vectors extending from subatomic particles of matter, a lack of distinction between energy vectors in matter and light has arisen that has led, in part, to the misguided attempts to include matter and light in a single unified field theory. (Light here is taken to include heat, radio and gamma rays.)

In the atom, electrons occur in pairs. Each member of the pair spins in the opposite direction. The relative directions of these motions are in parallel. They are not perpendicular so they can interact but because the interactions are in opposition they cancel each other out. The result is a cancellation of the magnetism. This is why most atoms are non-magnetic. Natural magnetism arises from atoms containing an unpaired electron. Because this magnetism is not cancelled it confers magnetism on the atom as a whole. Atoms of iron are magnetic because of the spin of an odd electron but iron only becomes magnetic when the atoms are aligned in the same direction. If they lose their alignment, through heating or hammering the iron loses its magnetism.

The representation of charge and magnetism as whirlpools and tops can help us to appreciate the way vortex vectors can be *layered*. Imagine the spin of water in a whirlpool as electric charge and the spin of a top as magnetism. Now imagine the whirlpool rotating in a tidal current. The motions are layered. In pyshics this illustrates the way vortex vectors can be layered. When additional motions are layered on vortices they set up additional fields of force.

When nothing exists but motion all forms and forces in the Universe consist of layers of motion interacting with each other and some of the interactions can be quite complex as one vector of motion layers on another.

The radial flow of concentric spheres of energy is the primary motion in the vortex, which gives rise to the radial vectors of electric charge. The vortex can then rotate. This secondary spin is called *quantum spin*. Quantum spin should not be confused with the quantum vortex. Quantum spin is a rotation layered on a quantum vortex. It is a perpendicular layer of motion that gives rise to natural magnetism; the *magnetic moment* of a particle. The rotating quantum vortex can then vibrate or move. These tertiary motions layered upon the primary and secondary motions of a quantum vortex can contribute to or cancel out force fields in the layers. The type of field depends on the relative directions of the overlapping and interacting energy so the interactions of the *layered vortex vectors* follow self-evident rules.

If layered vortex vectors move relative to one another along the radii of the vortices they interact as an electric field. If their relative movements are at a tangent to the vortex, they interact as a magnetic field.

The interaction rules between vortex vectors are simple and obvious. When vortex vectors move with each other in the same direction they attract. When they move against each other in the opposite direction they repel.

The vortex rules of interaction are demonstrated by electric charges. In positive charges concentric spheres of energy expand. When they overlap they move against each other in opposite direction as they grow out from their vortex centres. With negative charges the concentric spheres of energy contract. When they overlap they move against each other in opposite direction as they shrink into their vortex centres. The vortex vectors of like charges are effective in opposite direction so they repel. When positive and negative charges overlap, one set of vortex vectors are directed inwards while the other set are directed outwards. As the relative vectors of these opposite charges are effective in like direction, they attract.

The interaction rules of vortex vectors apply to magnetism as well as charge. There is a magnetic repulsion between two electrons spinning in same direction. This occurs because the rotating vortex vectors are in the opposite direction where they meet.

There is a magnetic attraction between two electrons spinning in opposite

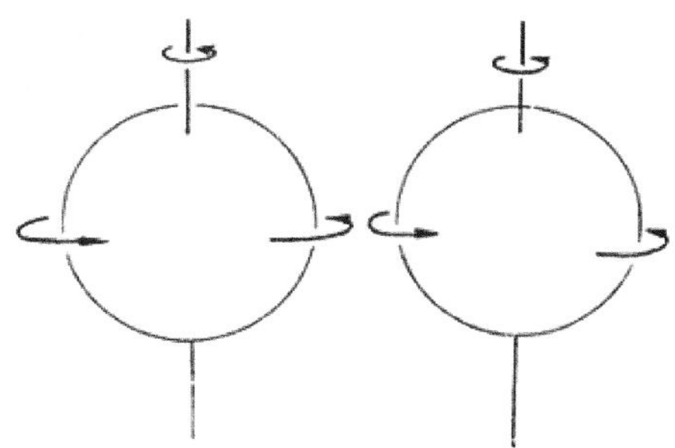

direction. This is because the rotating vortex vectors are in the same direction where they meet.

When electron vortices flow in an electric current there are vortex vectors

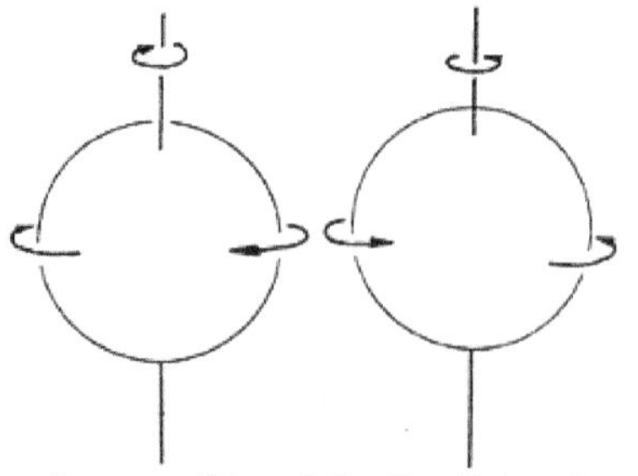

...because of the uni-directional interaction

of the rotating spheres of vortex energy

associated with this motion that are effective at a tangent to them. These set up concentric *rims of magnetism* that surround each moving vortex at right angles to the current of electricity. The concentric rims of magnetism would add up to form concentric cylinders of magnetism. Imagine slicing an onion then skewering it through its centre. In the line of the skewer, representing the current, the moving onion rings would circumscribe concentric cylinders.

If two current carrying conductors are laid side by side they will attract if the

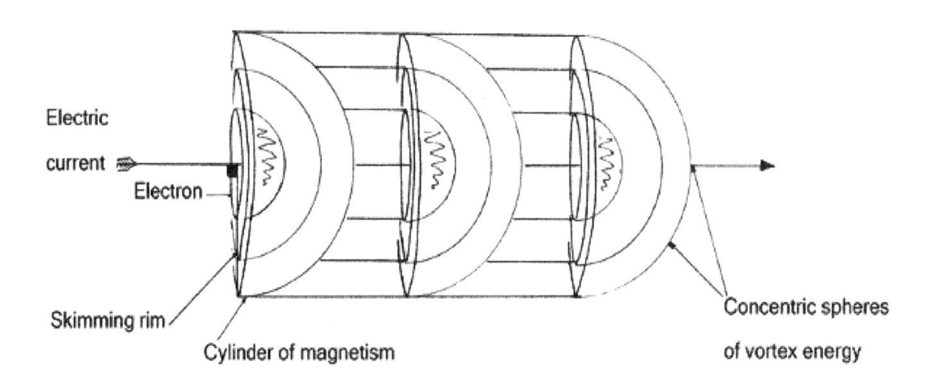

current is flowing in the same direction but repel if it is flowing in the opposite direction. This effect was first observed by the French scientist André Ampere.

When the concentric spheres of vortex energy, extending from the two

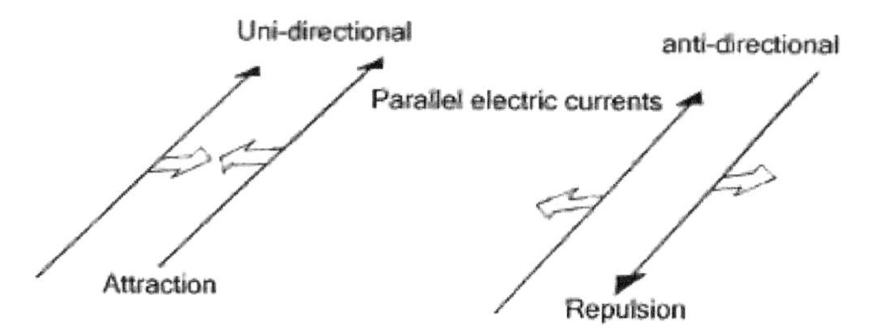

parallel electric currents, move against each other in opposite direction they repel. When they move with each other in the same direction they attract. The vectors are effective at a tangent to the concentric spheres of vortex energy so the interactions between the current carrying conductors are magnetic.

Michael Faraday discovered that a stationary magnetic field would not interact with a stationary electric field. This is because radial vortex vectors set up electric fields and tangential vortex vectors set up perpendicular magnetic fields. Because these fields are effective at right-angles they are distinct and non-interactive. But Faraday then found that a moving magnetic field can interact with an electric field. The movement of the magnetic field is a tertiary motion imposed on the primary motion of the vortex causing charge and the secondary motion causing magnetism. This new motion sets up a new set of vortex vectors. If they are radial in effect they act as an electric field. If they are tangential in effect they act as a magnetic field. Faraday discovered this effect when he realised that a moving magnetic field could act as an electric field and induce a flow of electricity in a wire. He used this discovery to invent a rudimentary dynamo which paved the way for electricity generators in power stations, motor vehicles and wind turbines.

Faraday also discovered that a growing or decaying magnetic field will interact with an electric field. In a growing or decaying magnetic field the intensity of vortex energy is changing. The intensity of energy in the spherical vortex is uniform over the surface of each concentric sphere whereas it varies along the radius. Uniform movement at a tangent to these spheres would not incur any change in intensity of energy. This is the characteristic of a magnetic field. However, movements along the radius of the vortex are associated with a change in intensity of energy. This is the characteristic of an electric field. Any change in intensity of vortex energy would be effective as an electric field and could interact with another electric field. A growing or decaying magnetic field will interact with an electric field because this change in intensity of energy is an electric effect. Alternating currents of electricity grow and decay many times a second. You can feel this as a buzz if you stroke an electric appliance. Growing and decaying magnetic fields are used to operate transformers that increase or reduce voltages. Transformers range from the mighty beasts in a national grid to the small chargers for mobiles, tablets and laptops.

In a Universe based on motion, vectors are crucial as they determine the relative directions of interaction between different forms of motion. The primary direction of energy in the quantum vortex is paramount. If the direction of primary motion in a subatomic vortex of energy is reversed, then this would affect any secondary motion superimposed upon it.

Particles with the same sign of charge, but spinning in opposite direction, experience a magnetic attraction. If the sign of charge of one of them is reversed then the magnetism will reverse and set up a force of repulsion. This effect has been observed between particles of matter and antimatter.

Just prior to annihilation electrons and positrons, undertake a 'death-dance' called *positronium*. If the electron and positron are rotating in the same direction they attract each other and consequently annihilate sooner. If electron and positron rotate in the opposite direction the magnetic repulsion delays their annihilation. It demonstrates an interesting effect of the layering of vortex interactions that reversing the sign of charge can affect a reversal of the direction of a magnetic interaction.

An account for electricity in the quantum vortex theory is based on the free movement of electrons in a conductive substance. When they are aligned by the application of a potential gradient along the conductor, movement can be transmitted from one electron to another, as compression and rarefaction vibration down the voltage gradient. Electricity is the result of these longitudinal vibrations of electrons in matter. To understand electricity, imagine electron vortices shunting against one other. Because of the infinite extension of vortex energy the shunt or longitudinal vibration would pass from one electron to another without them ever coming into direct contact. The shunt, or longitudinal vibration, is a radial vortex vector. Electricity is the passage of the shunt down the line of electrons rather than the passage of the electrons themselves. While the electrons drift slowly in the potential gradient at approximately three hundred kilometres per hour electricity travels at nearly the speed of light, about three hundred million metres per second.

CHAPTER 4
SPACE

I was very influenced by Albert Einstein. In 1974 I read a new biography on him entitled, *Einstein: His Life and Times.*[1] Apparently when he went to New York in 1919 a reporter asked him to put his theory of relativity in a single sentence. The great man replied, *"If you remove matter from the Universe you also remove space and time."*

Reading Einstein's remark confirmed something for me. I knew that if you move matter you also move space because they are the same thing. They are both vortex energy. When I read that book I had already concluded that space is an extension of vortex energy from matter into infinity. For me matter was the dense vortex energy we perceive and space was the sparse vortex energy extending beyond our direct perception. I concluded that matter and space are not different things, they are different ways we perceive the same thing. In 1970 I imagined the extension of every vortex of energy into infinity as a particle of space. I thought of it as a *quantum of space.* I premised the vortex energy, extending from every subatomic particle of matter in a body, as a *quantum field of space* surrounding the body. I envisioned everyone and everything surrounded by concentric *bubbles of space* which moved with them wherever they went.

The space bubble idea explained the central premise, in Einstein's special theory of relativity, that the *measured speed of light is independent of the velocity of the observer.* It was obvious to me why this would be so. If we measure the speed of light, the photons whose speed we are measuring would be in our own quantum field or bubble of space which would move with us as we move. Measuring the speed of light relative to our own space would cause the measure to be independent of our own movement. This is because our extension of space would be part of us. This would explain why the rotation of the earth doesn't affect flight times. When aircraft fly they would do so in the

1 Clerk R.W., Einstein: His Life and Times, Hodder & Stoughton, 1973.

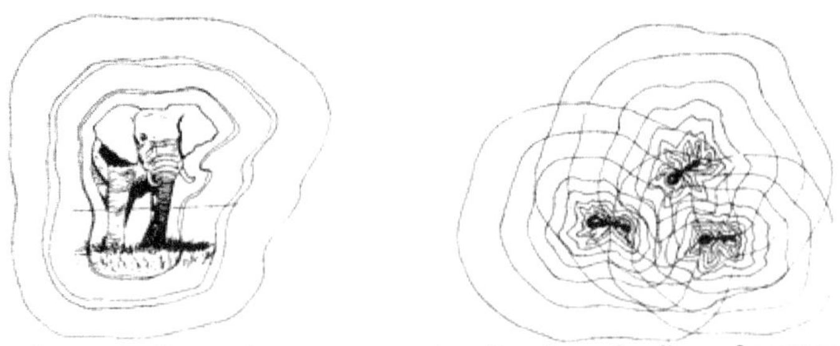

Space is the vortex energy extending from bodies of matter

Earth's extension of space which rotates with the Earth.

In 1887, Albert Michelson and Edward Morley carried out an experiment in Cleveland Ohio to see what effect the movement of the Earth would have on a measure of the velocity of light. They discovered the velocity of light was not affected by the velocity of the Earth. I realised this was because they were measuring the velocity of light relative to the Earth's bubble of space which, as an extension of the Earth, was moving with the Earth.

I got the realisation that space bubbles could explain Einstein's relativity from a dachshund called Johannes. In 1969, when the troubles broke out in Northern Ireland, I transferred to Queen Elizabeth College in London University, and

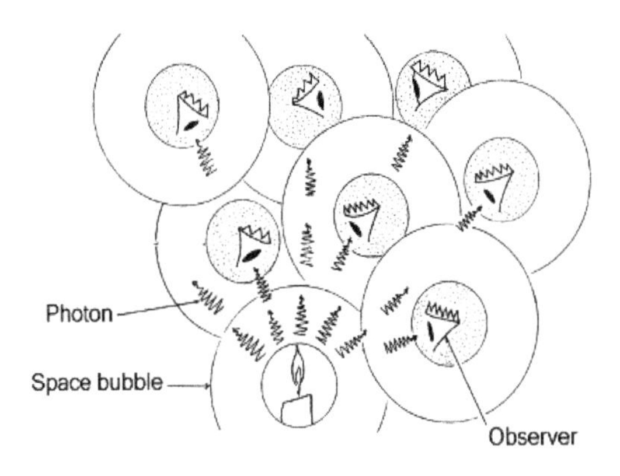

took up digs in Campden Hill Square, Kensington. The family I stayed with had a dachshund called Johannes. Most days I took him for a walk around the square. One day, in the spring of 1970, Johannes was unwell. He was wrapped in a blanket like a sausage roll to confine him to his bed in the front hall. As I burst through the front door a spectacle beheld me. Johannes, wrapped in his blanket, was off his bed and crawling across the floor toward me with his front paws. He wanted walkies but he couldn't pull himself out of his blanket roll because his blanket was moving with him as he moved across the floor. I had a eureka moment. I saw Johannes as a moving body and his blanket roll as the space extending from it. Watching him crawl across the floor in his blanket I realised a body couldn't move *relative* to its space if its space was moving with it.

As I thought about space, I realised *real* spheres of vortex energy would flow through *apparent* spheres of space. These were the concentric levels of intensity of energy set up by their distance from the centre of the vortex. I called them *shells of space* because they were not real. They were just the *fixed levels of intensity of energy*, in the vortex energy field, which set up the static, apparent forms of matter and space known to yogis as *maya*; the illusion of forms. Everything we think is real is unreal. Spin has us all totally deluded. Explaining my physics to my landlady, I used my large family as an analogy for this delusion. Almost every year a baby had been born into my family. As the babies grew into children and each one of us grew out of a year of age there was always another brother or sister growing behind us to take our place so that each year of age was always occupied by someone. The age years represented the shells of space and the family depicted a form of matter. The family was set up by the passage of real children through arbitrary years of age. When we all left home and went our separate ways that nuclear family unit no longer existed. The babies, in my analogy, depicted bubbles of energy popping out of the singularity point at the centre of a 'maternal vortex'. The child growing through each year of age corresponded to a real bubble of energy growing in size through each level of intensity, defined by its distance from the centre of the vortex. These levels of intensity of energy, setting up the forms of matter and the shells of space, persisted not because they were real but because there was always a sphere of vortex energy at each level of intensity. I got this idea from watching the flow of water through vortices and ripples in a stream. The water was real but the vortices and ripples were not. They were transient forms that lasted only while water flowed through them. In like manner the levels of intensity of vortex energy are not real things. The seemingly static nature of space and matter, set up by vortices of energy, give us an illusion of substantiality that deludes us into imagining the unreal as real.

Space is electrically neutral. This could be because there are an equal number of oppositely charged vortex particles in existence which extend into infinity and cancel out each other's electric effects. As the spheres of vortex energy extend into infinity their intensity would drop but the diminishing intensity of energy from every vortex would be compensated for by the addition of vortex energy extending from other bodies. In that way universal space would be maintained as a collective of the *quantum fields of space* extending from each and every vortex of energy in existence.

The quantum principle of the whole being made up of the parts, which applies to every form of energy, can be applied to space but only if we break from the opinion of Democritus and stop thinking of space as a void. Einstein contradicted Democritus when he proposed space is a real form of energy that moves with matter and can be distorted by mass. I was in awe of Einstein but I didn't always agree with him. In my vortex theory I contradicted him on a number of occasions. I agreed with him that space is a real form of energy but when Einstein said space is distorted by mass I said no. I contended that space is shaped not because it is distorted by mass but because it is an extension of mass. As an extension of matter the shape of space would be an extension of the shape of the massive body it extends from.

Extending from the quantum vortex, the Earth and Sun and all the other planets and stars in existence, the concentric spheres of vortex energy would set up interlacing spheres in the vastness of space. I also realised if a vortex of energy sets up space then the subatomic particle, as a quantum vortex, could be described as a three-dimensional spiral space path. I went on to use that idea to explain nuclear physics.

CHAPTER 5
NUCLEAR PHYSICS

The theory of the quantum vortex provides a straightforward account for nuclear energy and high energy physics from the ideas that subatomic particles are spiral space paths and that naturally occurring subatomic vortex particles are very stable. The stability of the natural quantum vortex is evident from the stability of the Universe.

As a lively lad learning in my grandfather's workshop, I twisted a rod of steel through a spiral die and watched it come out the other side as a screw. That experience prepared my young mind for the way I would use the vortex theory to explain how unstable particles of matter can be synthesized in high energy laboratories The idea I had was that the short-lived particles that appeared in cosmic ray and high energy lab experiments were vortices with *transient mass* generated by forcing energy through spiral space paths in stable nuclear vortices as though they were *vortex dies.*

I used an analogy of doughnuts to explain what was happening in the production of unstable vortices. Batter takes on the shape of a doughnut when it is forced through a doughnut mold but if the doughnuts were allowed to drop onto the floor very soon after leaving the mold they would cease to be doughnuts and revert to a mess of batter.

In the quantum vortex theory, the idea I had was that the particles generated in high energy experiments were formed by forcing wave kinetic energy through natural vortices of energy. My idea was that in its passage through a natural, quantum vortex the kinetic energy would take on the form of a vortex and as it spirals through the stable quantum vortex the unstable vortex would contribute to its mass. This mass would be transient as on emerging from the natural quantum vortex the unstable vortex would revert immediately to its natural wave kinetic form again. The *pi-meson* particles that emerged from the nucleus of the silver atom, in Cecil Powell's cosmic ray experiment, were an example of this.

Analysis of Powell's photographic plate[1] showed that the mesons, appearing from the nucleus of the silver atom, lost their mass in the time that it takes for a quantum of energy, traveling at the speed of light, to traverse an atomic nucleus. The time amounted to 10^{-15} sec.

To account for the formation of the short-lived particles of matter in Cecil Powell's experiment, and subsequent experiments in high energy particle accelerators, like CERN and Fermilab, I proposed two simple and self-evident quantum laws of motion.

Quantum Law of Motion I:	*Particles of energy will maintain their original form of wave or vortex motion unless a change of form is forced upon them.*
Quantum Law of Motion II:	*If the force of change of is removed the particle of energy will revert to its original form of motion.*

Springs are the model I use for the quantum laws of motion. A change of form can be imposed on a spring but if the change is no longer imposed the spring will revert to its original form.

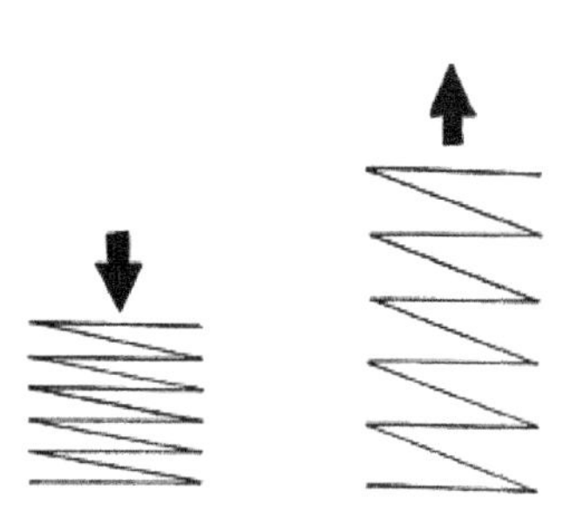

According to the first quantum law of motion if a quantum of waves drives through a quantum vortex of energy it would take on the form of the vortex and contribute to its mass. According to the second quantum law of motion if the quantum of waves escapes from the vortex the contributed mass would be lost and as it reverts to its original kinetic state, as a train of propagating waves, it would radiate away.

As physicists increased the energy in their experiments, they produced ever more massive particles. This was because the increasing kinetic energy driving through the natural vortices in atomic nuclei was increasing the mass produced from the transformation of wave kinetic energy into static vortex energy. The short life of the new particles upholds the quantum laws of motion.

However, in the high energy laboratories physicists were discovering particles that were lasting considerably longer. The difference in lifespan was a thousand-fold. Instead of lasting for a trillionth of a second they were surviving for a billionth of a second.

Murray Gell-Mann, who invented the quark theory to explain the outcomes

1 McKenzie A. E., A Second MKS Course in Electricity, Cambridge University Press, Plate 19

of high energy physics, called these new particles strange for their exceptional longevity. He speculated the existence of additional quarks to account for them and suggested arbitrarily that this was a hitherto unknown property of matter, which he called it *Strangeness*. I accounted for Strangeness with the quantum laws of motion.

My account was a development of the original axioms of the vortex theory. I included no additional arbitrary elements so I could invoke *Ockham's razor* to claim the quantum vortex is a better theory than Gell Mann's quark theory.

The 14th century philosopher William of Ockham is famous for stating: *"non sunt multiplicanda entia praeter neccessitatem* - Entities are not to be multiplied beyond necessity." He argued that in a choice of theories we should accept as true the one that has the least arbitrary assumptions. He argued this law of economy of ideas with such sharpness that it became known as Ockham's razor.

When strange particles decay they leave a proton or an electron behind. My suggestion is that when kinetic energy drives through an atomic nucleus, in a high energy experiment, an unstable vortex sometimes forms around a stable electron or proton vortex in the nucleus. Should this occur the electron or proton vortex would have a *scaffolding effect* on the unstable vortex, conferring upon it a degree of stability that increases its lifespan. The scaffolding effect could stabilise the change of form from wave to vortex in accordance with the first quantum law of motion. In so doing it would delay the action of the second quantum law of motion by slowing down the reversion of swirling energy to its natural wave-kinetic form. There are no additions to my theory to account for strange particles. They merely represent a delay in action of the second quantum law of motion.

As physicists increased the energy in their accelerators, they found ever more massive particles were generated in layers around a natural particle. I imagined this occurring much as ice forms in layers around a particle of dust

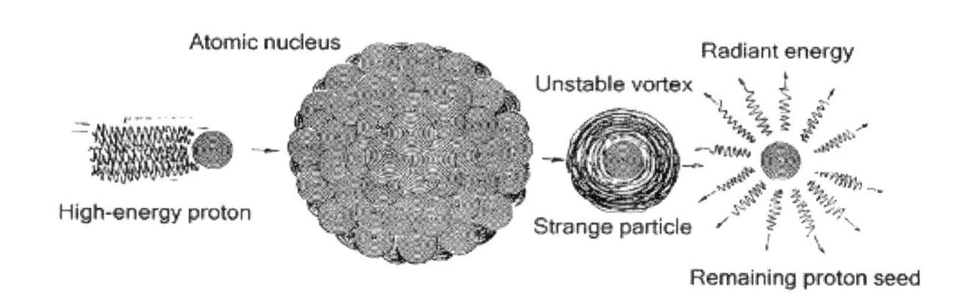

in a hailstone.

The analogy I used to explain how some of the newly formed massive particles decay in steps, or layers, to reveal a new, lighter particle at each lower level of energy, was a *hailstone.* This *cascade decay* of unstable particles fits with the model of a hailstone melting layer by layer, in reverse of the way it formed. The levels of mass-energy forming round a proton in the creation of a strange particle are called *doses of strangeness.* The decay of the strange particles in steps, to reveal lighter particles on each step, is described as *shedding doses of strangeness.*

According to the vortex theory, as the cascade decay occurred and each dose of strangeness was shed it left behind a smaller less massive vortex particle. The scaffolding effect of the proton vortex over the residual transitional mass would therefore be stronger because there would be less swirling energy to stabilise also it would be closer to the stabilising proton vortex at the centre of the swirl. This increase in the strength of the scaffolding effect meant that as each dose of strangeness was shed the particle left behind would last longer than the one that went before it. This is evident in the cascade decay of a particle called the *Omega-minus.* In 1964, at the Brookhaven National Laboratory, Nicholas Samios photographed a cascade decay the Omega minus.[2]

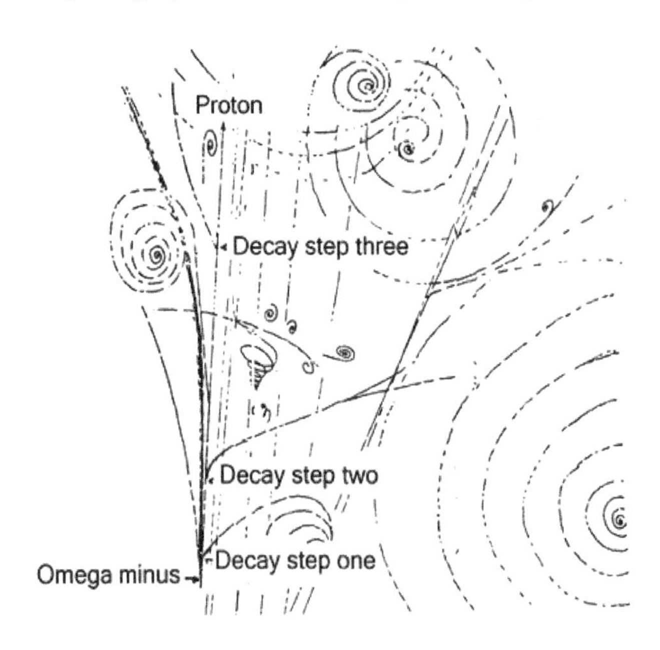

If you look closely at the image of the three-step cascade decay of the Omega

2 Calder N, Key to the Universe: A Report on the New Physics, BBC Publications 1977

minus particle, you will notice that the distance between each successive step in the progressive cascade decay is longer. The length of the track a particle leaves before decaying gives a measure of its lifespan. This shows that each residual particle, in the cascade decay process, lasts longer than the particle that went before it.

I went on to develop an account for nuclear energy, and also the nuclear binding force, in a branch of the quantum vortex theory I called *Capture theory.* Capture theory came from a realisation that over the vast periods of time they have been in existence protons would have become saturated with captured kinetic energy. This would occur if protons were quantum vortex particles because these would be, in effect, increasingly tight spiral space paths. My idea was that on encountering the spiral space path presented by a proton quantum vortex a wave-train quantum of energy would drive into it and become captured. This form of *natural capture* would occur in line with the first quantum law of motion.

Particles of kinetic energy driving into the progressively tight spiral of a proton vortex would be trapped by virtue of its spiral shape. The particles of kinetic energy would drive into the *static inertia* of vortex motion and transform themselves into the mass of transient vortex energy by virtue of their *kinetic inertia.* The proton vortices would be passive in this process. I used crab pots as a model for this natural process of energy capture.

I visualised the static quantum vortex as a crab pot and the kinetic quantum as a crab. Crabs crawl into sedentary pots and are caught by virtue of the shape of the pot. The pots take no active role in the process of capture. In like manner the wave-train quantum of vibrating energy would drive into a vortex of energy and be captured by virtue of the vortex shape forming a spiral space path. The vortex, with its static inertia, would remain passive and not take an active role in the capture process.

In the vortex theory the diameter of a proton is defined as the limit of its ability to *completely capture* energy. This would be a reflection of the tightness of its three-dimensional spiral.

Imagining proton vortices were saturated with the energy they had captured, I considered captured energy to be responsible for the *meson* mass that they contained. The mass of the pi-meson or *pion* that appeared in Powell's experiment represents the capacity of a proton to contain captured energy.

In line with the first quantum law of motion as wave-train particles of kinetic energy are captured by a vortex of energy they would be changed into vortex motion and so transformed into mass. In line with the second quantum law of motion if they escape from the vortex they would lose their mass as they revert from spin to their original wave form.

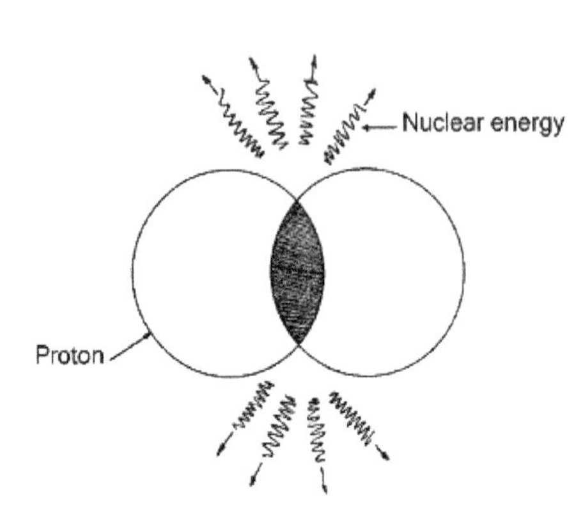

In the high temperatures of the sun, or an exploding hydrogen bomb, protons and neutrons have sufficient kinetic energy to attain very high velocities and collide with an immense force. When they do so they converge. If proton or neutron vortices collide and converge their capacity to contain captured energy would be reduced so some of the captured energy swirling inside them would be lost as radiant energy.

In accordance with the second quantum law of motion the transient mass of vortex energy escaping from the proton or neutron vortex would immediately revert to the original kinetic form and radiate away. This is how *transient mass* would be lost as radiant energy in nuclear fusion. I used this idea to account for *nuclear energy* in the vortex theory.

Buckets of water helped me describe nuclear energy. Protons or neutrons saturated with captured energy could be likened to buckets full of water. If one bucket of water is placed inside another its capacity to contain water would be reduced so some of the water would spill out. In like manner as protons or neutrons converge their capacity to contain captured energy would be reduced so some of the captured energy would spill out as nuclear energy.

Captured vortex energy, as transient mass, contributes to the mass of a proton or neutron so the loss of captured energy would be associated with a loss of the overall mass of the proton or neutron. In nuclear fusion 0.7% of the mass of the combining protons or neutrons is lost. This loss of mass, which is a loss of captured energy, accounts for the release of nuclear energy that occurs as hydrogen nuclei fuse together to form helium in the sun or in a hydrogen bomb.

Richard Feynman said *"Nuclear energy...we have the formulas for that, but we do not have the fundamental laws. We know it's not electrical, not gravitational and not purely chemical, but we do not know what it is."* [2]

In *Hidden Journey*[3] Andrew Harvey quoted an Oxford academic as saying, *"Only scientific criteria for truth are valuable and mystics are pathological cases."*

The simple and straight forward account for nuclear energy, provided in these few pages, comes from the insight of mystics who peered into the atom with their minds rather than machines and saw that the smallest particles of matter were vortices of energy. In the pre-scientific era yogi mystics probed matter with supernormal powers called *siddhis*. This practice by yogis was recorded in about 400 B.C.E., in the Yoga Sutras of Patanjali where the results of meditation were described in detail. In Aphorism 3.46 it states that through meditation the yogi can gain an extended faculty of observation from the practice of the *anima siddhi* - an ability developed by yogis to shrink their consciousness commensurate with the very small.

Using the anima siddhi mystics in the tradition of yoga investigated the smallest realms of matter with their minds and actually saw vortices of energy therein. In so doing they appear to have elucidated the holy grail of science. If mystics saw how energy forms mass and did so thousands of years before Einstein; if mystic perception has led to a unified account of practically everything in physics, including nuclear energy; if a substantial breakthrough in science can come through pre-scientific mysticism, then perhaps it is time for academics to reconsider their attitudes towards mystics.

Only a small amount of captured energy is lost as nuclear energy in nuclear fusion. Most of it is left behind spinning inside the converged vortex particles. I presumed it could swirl between them after they converged. If this were so the residual captured energy might bind them together.

Captured energy, swirling between nuclear vortex particles could fuse them together. This would explain the *strong nuclear force*, which binds protons and neutrons together in the nucleus of an atom. This simple account for nuclear binding explains why it is short ranged, does not have polarity or infinite extension and is limited to the diameter the atomic nucleus.

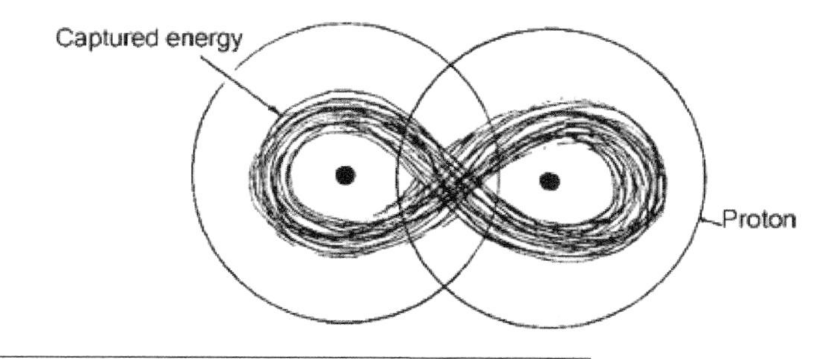

3 Harvey A. Hidden Journey Rider, 1991.

If the captured energy shared between quantum vortices in the nucleus of an atom is responsible for nuclear binding and corresponds to the pi-mesons in protons and neutrons, that would fit with the Yukawa prediction of mesons being responsible for the strong nuclear force. The vortex theory for nuclear binding can also explain why the binding between nuclear particles increase as the binding energy decreases. Let me explain.

Uranium and plutonium have large nuclei. When these divide during *nuclear fission* they split into smaller nuclei. In the smaller nuclei the proton and neutron – *nucleon* - particles bind together more tightly. This occurs because in smaller atomic nuclei the distances between the centers of the nucleons are less than they were in the large uranium or plutonium nuclei. The vortex account for nuclear fission is that as the nucleon vortices clamp together more tightly, after fission, in the formation of smaller nuclei they squeeze out a small amount of the captured energy. This is lost as nuclear energy. Meanwhile the majority of the captured energy, left behind swirling between the nuclear vortex particles, follows a shorter circuit in the smaller nuclei. That would cause the nuclear vortices to bind more tightly together. After nuclear fission the loss of a minute amount of binding energy would be marginal compared to the increase in binding between nuclear particles due to the large amount of residual captured energy racing in a tighter circuit between them. That would explain why, after fission, binding increases in the residual nuclei despite a loss of binding energy.

I use a hedgehog analogy to help explain nuclear energy and nuclear binding. Imagine the proton as a hedgehog and the captured energy as its fleas. In much the same way that a hedgehog carries a resident population of fleas so a proton would carry a population of captured particles of kinetic energy.

Hedgehogs don't do anything to acquire fleas, the fleas just hop onto them. So with a proton it does nothing to acquire its captured energy. It is the quantum wave-trains of kinetic, vibrating, energy that drive into the spiral space of the proton vortex.

If a hedgehog is weighed, the greater part of the mass would be that of the hedgehog. A lesser part would be that of the fleas. In like manner the greater

part of the measured mass of a proton would be that of the stable proton vortex but a lesser part would be that of the unstable swirl of captured energy it contains.

The hedgehog prickles represent the charge repulsion between protons. Because of their prickles hedgehogs converge with difficulty. So it is because of their charge repulsion, protons converge only if they collide with considerable force.

Imagine if two hedgehogs were forced together, as their prickles converged there would be less space between them for fleas and so some of the fleas would be evicted and hop away. In like manner as two protons converge with force there would be less space within them for captured energy and so some would be lost and would radiate away as nuclear energy.

Because of the loss of some of their fleas the mass of the converged hedgehogs would be less than the sum of their weights before they were pushed together. So it is the mass of two converged protons would be less than the sum of their masses before they collided.

Most of the fleas would remain on the hedgehogs and not being bothered about which back they bite they would hop from hog to hog. In like manner captured energy belonging to the proton vortices would circulate between the converged protons and by this action would bind them together.

When asked if my vortices are quarks I reply, *"I don't believe in quarks. 'From my perspective quark theory is unnecessary because it was developed to account for the results of the high energy experiments which I explain with my quantum laws of motion."* Let me share another reason why I don't believe in quarks.

On the 27th January 1977, BBC televised a report on the new physics in a programme entitled, The Key to the Universe. In a companion book under the same title,[2] Nigel Calder wrote about how the theory for quarks originated: In the early 1930 the contents of the Universe seemed simple. From just three kinds of particles, electrons, protons and neutrons you could make every material object known at the time. Thirty years later human beings were confronted with a bewildering jumble of dozens of heavy, apparently elementary particles, mostly very short lived. They came to light either in the cosmic rays or in experiments with the accelerators. The particles had various mass-energies and differing qualities such as electric charge, spin, lifetimes and so forth. Moreover, they were given confusing, mostly Greek names so that one of the most eminent of physicists, Enrico Fermi, was driven to remark before his death in 1954,"If I could remember the names of all these particles I would have been a botanist."

The proliferation could be understood, to some extent, in that many of the particles seemed to be energetic relatives of the proton. Because they possessed greater inherent energy their masses were greater. Each was in some sense less tightly bound together than a proton and it could quickly change into a proton with a release of binding energy and an associated loss of mass. But that implied the proton was not a truly basic particle: it was made of something else, which could be bound more or less tightly together.

A small group of theorists brought order out of chaos. The principal figure amongst them was Murray Gell-Mann of Caltech (The California Institute of Technology), then in his early thirties. He declared that all the heavy particles of nature were made out of three kinds of quarks. He had the word from a phrase of James Joyce 'Three quarks for Muster Mark.' It was the mocking cry of gulls, which Gell-Mann took as referring to quarts of beer, so he pronounced quark to rhyme with 'stork'. Many other physicists rhymed it with 'Mark'. In German, as sceptics were not slow to notice, 'quark' was slang for nonsense....

There is a lot in a name and in my opinion the quark theory is nonsense, as the name suggests, not because of mocking gulls but because of a black swan.

The Austrian philosopher, Karl Popper, used an analogy of black and white swans to explain that science is not in the business of proving theories but rather of disproving them.[4] Someone could have a theory that all swans are white but even if a hundred white swans were counted, their belief would not be proved and the addition of more white swans wouldn't make it any

4 Popper K. Logic of Scientific Discovery, Hutchinson 1968

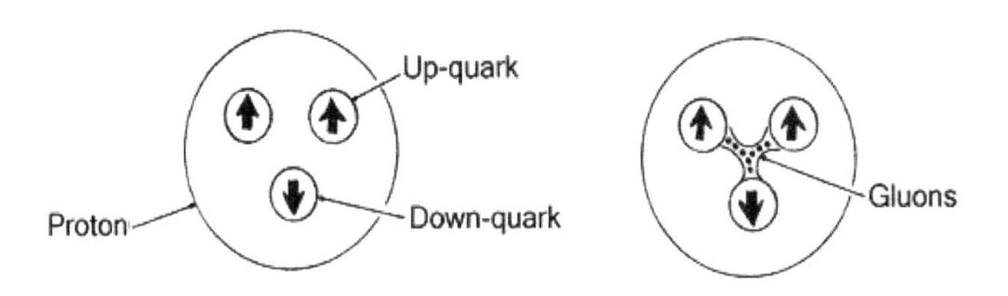

truer but the appearance of a single black swan would disprove the theory altogether. Scientific theories are white swan belief systems, which survive until they are destroyed by the appearance of a black swan, that is the arrival of even a single fact which makes it clear that theory cannot be true. The black swan for the quark theory is the proton. My nearest town is Dawlish. Its emblem is the black swan. I love to walk by the stream running through the centre of town and watch the black swans. They remind me of the knock-out blow to quark theory.

The single fact which throws quark theory into question concerns the lifespan of a proton. The life of a proton has been estimated at 10^{33} years - that is a billion, trillion, trillion years - whereas one ten billionth of a second is considered a strangely long-lifespan for any of the new particles found in high-energy research. As spontaneous proton decays have never been observed the proton is infinitely more stable than any of the new heavy particles that have come to light in the high energy laboratories.

Physicists call the heavy particles *baryons* from the Greek word for heavy. They account for the difference in lifespan between protons and other baryons in a law called the law of *Conservation of Baryon Numbers*. However, this law doesn't tell us why protons live for so much longer than the other particles, it simply states mathematically that they do. It is just one of the many instances in science where a statement is made that phenomena are observed to behave in a certain way, but no reason is given for why that is so.

With over 31 million seconds in a year the difference between the estimated lifespan of a proton - in the order of 10^{38} seconds - and one of the new particles lasting in the order of 10^{-15} seconds is a figure in excess of 10^{50} (a number with fifty zeros behind it). Mathematicians use the number 10^{50} as a cut off point for probability. If the probability of something happening is less than 1 in 10^{50} then the convention is it never occurs.

With a difference in stability in excess of 10^{50} it can be said that as it is highly improbable protons are of the same order as other baryons. There must be a fundamental difference between them and the baryons appearing in high energy experiments to account for the difference in their stability. Protons cannot be lumped with the unstable baryons unless reason is found to account for the disparity in their lifespans.

This point is obvious from a simple analogy of building sites. Imagine walking down a road between two building sites. On the site to the left the houses disintegrate as soon as they are built. Within a trillionth of a second of the last lick of paint being applied they vanish. On the site to the right the houses are advertised for sale with a trillion-year guarantee. It would be insane to assume identical bricks, mortar and construction techniques are employed on both building sites without providing an account for the difference in the durability of the houses. The baryon particles are like houses in this analogy. The different types of quark are equivalent to bricks and the gluon bonds binding them together are likened to mortar. Those who believe in quark theory cannot claim to be sane let alone scientific if they accept different combinations of quark bricks cemented with gluon bonds could give rise to all the baryons, including protons, with no account given for why protons are infinitely more durable than all the other baryons.

Quarks could be used to explain protons or they could be used to account for the other baryons but not both. To use quark theory to explain only protons would be pointless because the theory was invented to explain the new baryons. But if quark theory provided an explanation for all the baryons apart from the proton it would still be pointless because much of the observed Universe consists of protons whereas all other baryons, apart from the neutron, have been observed only in high-energy research. If the theory were to exclude protons high-energy research would be pointless. This is because physicists may talk about discovering new particles of matter in the high-energy research but to say they have been discovered is misleading because discovery implies that something already exists and is waiting to be found. The short-lived baryons have been created out of the massive amounts of energy fed into the high-energy particle accelerators. These new heavy particles synthesized in the accelerators have no place in normal matter. They are just anomalies of high-energy experiments. For example when in April 1994 physicists at Fermilab in Illinois claimed to have discovered the top quark they didn't actually see a quark. What they saw was the tracks of a jet of electrons and muons, which they supposed to be the breakdown products of W-particles, which were then assumed to be remnants of a top quark. The discovery was seen as a triumph and the link between quark theory and Democritus' materialism was trumpeted in the media. In the May 15th *Star Tribune* in Minneapolis, Jim Dawson wrote, *"So there we have it, after more than two thousand years of searching, all of the fundamental stuff of Democritus' atom has been revealed. The crowning moment came a couple of weeks ago, when physicists announced that a gigantic, 5,000-ton machine apparently had detected a very small particle called the top-quark."* [5]

Vast sums of money have been spent on building and running massive particle accelerators. In these cathedrals of Democritism scientists created a host of new particles which were explained by quark theory. The theory threw up new difficulties and exciting predictions which required more research and an

5 Dawson J, Star Tribune of Minneapolis, May 15th 1994

ongoing need for ever more powerful accelerators. Everything in the world is supposed to be made of quarks and yet, though no expense or effort has been spared in the effort to find one, not a single quark has ever been observed in a free state. So why do physicists believe in quarks?

Quark theory became generally accepted in 1968 after a series of experiments at the Stanford Linear Accelerator in California (SLAC) where electrons were accelerated down a three-kilometre long 'vacuum tube' by intense radio pulses and then targeted on protons in liquid hydrogen. The results of these experiments showed that electrons were being scattered - that is bounced back - from what appeared to be something small and hard within the protons in the hydrogen atoms. This inferred that the protons were not truly fundamental but contained smaller particles. Physicists were looking for quarks and naturally concluded that their bombarding particles were bouncing off quarks in the proton. These experiments caused the quark theory to become generally accepted.

The results of the SLAC experiments could be accounted for in another way. The meson captured energy swirling inside the proton could set up the apparent hardness in the proton. To understand this imagine a quantum vortex as candy floss. Candy floss is spongy due to the space between the strands of sugar. Likewise the spiral space within a vortex should make it spongy. However, the proton vortex would not be spongy because its internal space would not be empty. Filled with captured energy the proton would be more like a hard candy ball gobstopper than spongy candy floss.

Someone may develop a vortex theory for quarks but as far as I am concerned quark theory is just a deeper dig into the mines of materialism. Quark theory is complex and full of arbitrary assumptions. Quarks would add an unnecessary level of complexity to the quantum vortex theory. As far as I am concerned quarks do not exist. They are a figment of Murray Gell-Mann's imagination. In my humble opinion, electrons, protons and neutrons, along with their antimatter particles, are the real, natural subatomic particles.

CHAPTER 6
QUANTUM VORTEX MECHANICS

Early in the 20th century physicists discovered that particles can behave as waves and waves as particles. This came to be known as *wave-particle duality.* In the vortex theory wave-particle duality is explained in terms of the interactions between quantum wave particles and quantum vortex particles, described here as *wave-vortex duality.*

Quantum vortex mechanics results from a wave train particle of energy driving into the spiral space of a subatomic quantum vortex of energy where the wave-kinetic quantum is only partially captured. The uncaptured part sticking out of it pushes it forward in wave motion. This sets up a wave-vortex duality of vortex and wave particles of energy.

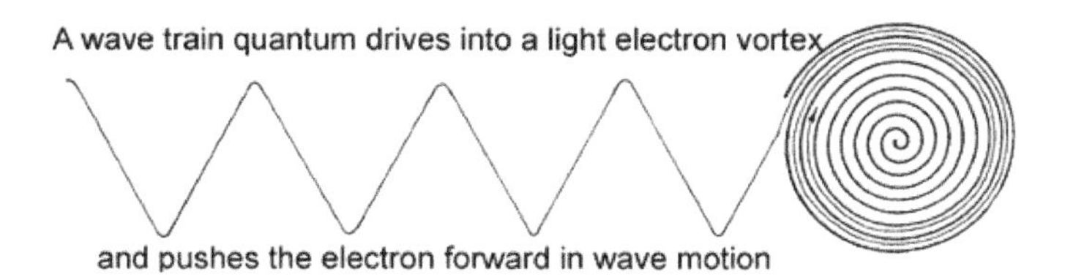

Tadpoles provide a model for wave-vortex duality. Just as the wavy tail of a tadpole drives the tadpole forward in wave motion so a wave-quantum caught by a quantum vortex of energy could drive it forward in wave motion.

Another analogy for wave-vortex duality is the capture of a monkey by a jar of nuts. The monkey represents a wave train of energy and the jar of nuts represents an electron vortex. The monkey puts its hand in the jar to get the nuts, but with a fistful of nuts it can't withdraw its hand so it is effectively caught by the jar. The jar is passive in the process. That illustrates the passive role of quantum vortex in energy capture. The greedy monkey depicts the quantum wave-train that takes the action of driving into it. Rather than let go of the nuts the monkey runs off with the jar attached to it so the jar takes on the mobility of the monkey. That depicts the electron vortex taking on the active inertia of the kinetic energy it has captured. The jar is heavier for the hand of the monkey contains. That represents the transient mass of captured energy swirling inside the electron vortex.

Quantum vortex mechanics is mostly concerned with wave-vortex duality and is more dedicated to electron vortices than proton vortices as electrons are less massive than protons and so are more easily driven into wave motion. With its greater mass and static inertia the proton vortex is more likely to withstand the impact of a quantum. That is why a quantum wave train of energy can drive right into it and become completely captured.

The mass of an electron is nearly two thousand times less than the mass of a proton so its static inertia is insufficient to withstand the kinetic inertia of an impacting quantum of energy. As the quantum of wave-kinetic energy drives into the spiral space path of the electron vortex and the electron moves forward under the impact only a part of it is captured. I call this *partial capture*. Imagine most of the quantum sticking out of the electron vortex like a tail, because the quantum is nothing but a particle of movement the wave-quantum could not cease to move in a wavy line. In consequence this *partially captured* quantum would drive the electron forever forward in wave motion. This would explain why electrons are constantly on the move, why they behave like waves and why it is difficult to know where they are from one moment to the next.

The radius of an electron at 2.81×10^{-15} metre is not a boundary to the electron vortex. It is the extent of its ability to *partially capture energy as mass*. It is the transient mass filling the space of the electron vortex that defines it as a corpuscular particle. I call the ability of captured energy to circumscribe the vortex, as a corpuscular particle with a measurable diameter, the *gobstopper effect.*

The proton radius at 0.84×10^{-15} is not a limit to the extension of the proton vortex. It is the limit of the proton vortex to *completely capture energy as mass.* Again this is the gobstopper effect of captured energy filling the internal space of the central regions of a quantum vortex giving it corpuscular definition and apparent hardness.

To help explain the difference between complete and partial capture I use a gate and post analogy. Imagine a proton as a gate post and an electron as a swinging gate. A nail, representing a wave-quantum, is hammered into the post. The post doesn't move so the nail goes right into it. The nail has been *completely captured* by the post. Now, an attempt is then made to hammer a nail into the swinging gate. As the gate moves under the impact of the nail only the tip of the nail goes into it. The nail has been only *partially captured* by the gate but the tip of the nail in the gate contributes to its mass.

The different types of capture reflect the different degrees of static inertia in the post and the swinging gate. If an attempt is made with a sledge hammer to drive a big nail into thc post, the post might move under the impact of the hammer blows. Likewise, if a proton saturated with completely captured energy is impacted by a high frequency quantum, with sufficient kinetic inertia to overcome its static inertia, the quantum could become partially captured by the proton. Though weakly bonded the wave-kinetic-quantum could drive the proton forward in wave motion. Applied to atoms, partially captured energy could account for *kinetics* which is concerned with increased vibrations in matter, measured as a temperature rise, when matter is heated.

In terms of the first quantum law of motion if part of a quantum drives into a quantum vortex and is partially captured it will contribute to the mass of the vortex of energy. This is why electrons and protons are more massive when they are moving than when they are at rest.

Returning to inertia. The inertia of the kinetic quantum of waves is to keep going unless stopped and the inertia of the quantum vortex is to stay put unless it is moved. Wave-vortex duality is a tussle between these opposing inertias.

The quantum laws of motion can be stated in terms of inertia. The two types of motion - wave propagation or spin - set up the opposite types of inertia.

Quantum Law of Inertia I:	***Particles of energy will maintain their original form of wave-kinetic or vortex-static inertia unless a change of inertia is imposed upon them.***
Quantum Law of Inertia II:	***If a change of inertia is no longer imposed the particle of energy will revert to its original form of inertia.***

At a quantum level the Universe operates on the interaction and balance between the opposite inertias. *Kinetics* and *work,* in physics, would be terms applied to an operation of the first quantum law of inertia, when waves of energy interact with vortices of energy. An example of this is water heating up. In physics the term *entropy* would apply to an operation of the second quantum law of inertia, when waves of energy cease to interact with vortices of energy. An example of this is water cooling down.

To recap, quantum wave-particles and quantum vortex-particles act in opposition so their interactions can be described in terms of opposing inertias. As the kinetic inertia of the wave form of energy pushes the vortex form forward in wave motion the static inertia of the ball vortex resists the attempt to be pushed forward.

In partial capture, however, the opposite inertias are *shared.* Shared inertias set up a bonded state between a wave quantum and a quantum vortex. This occurs as the vortex assumes the kinetic inertia of the wave quantum and the wave quantum shares the static inertia of the vortex.

An initial quantum of energy, partially captured by an electron, would be caught in the innermost, tight spiral space path in the electron vortex. This sets up a very stable bond between that quantum and the electron vortex. This strong wave-vortex bond, responsible for their wave-vortex duality, is exemplified by the *ground state* electron in an atom. The ground state electron is an electron in its normal orbit in an atom.

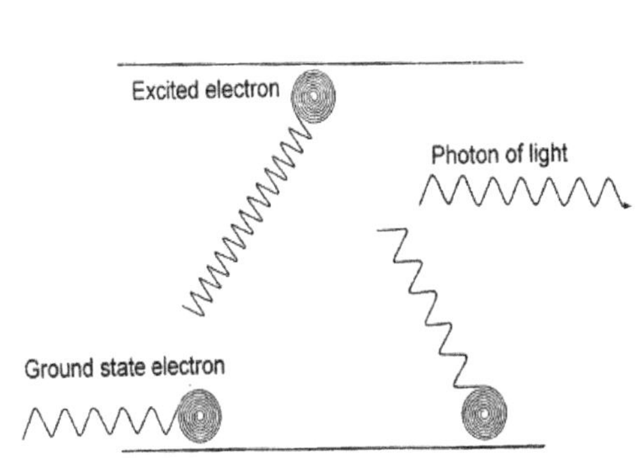

However, an electron can partially capture additional quanta of energy. Though the bonds are less stable, these can move it into a *higher quantum state.* If an electron absorbs a quantum with sufficient energy - represented by its frequency - to shift it into a higher orbit it can take a quantum leap. When a ground state electron leaps into a higher orbit this is called its *excited state.*

In terms of the vortex theory the excited state of an electron is less stable than the ground state. This is because the quantum driving the ground state occupies the central tightest regions of spiral in the electron vortex, whereas the quantum setting up the excited state would be captured more peripherally in the less tight regions of spiral in the electron vortex and so it would be weaklier bonded to the electron. As a consequence it could break free more easily and escape partial capture - in line with the second quantum law of motion. Then it could be released as a photon of light. This would explain *photon emissions* where the frequency of the photon emitted is equal to the difference in energy between the ground state and excited state orbits in the atom. The electron, having lost energy, would revert to its original ground state again.

If you strike a match the heat released by the combustion will excite electrons in carbon atoms floating in the hot air currents rising from the burn. As excited electrons leap to a higher orbit then drop back to the ground state again they release energy as photons of light. This sets up a *flame* and the frequency of the yellow light of the photons coming from the flame represents the difference in frequency between the ground state and excited state in the carbon atoms that absorbed energy from the combustion of the match.

The partial capture of wave kinetic energy, by electron particles of vortex energy, can help us understand chemistry. In the combustion of the match electrons in atoms of carbon in the wood get excited and leap across into the orbits of oxygen atoms in the air. The electrons then move between the atoms of carbon and oxygen. By this action the electrons bind the atoms together to form carbon dioxide. In chemistry this is called *covalent bonding.*

In the ignition of a match head some of the atoms form salts in consequence of the combustion. In this situation excited electron vortices, driven by partially captured energy, leave their orbits in one atom to move permanently into orbit in another. The two atoms, by gaining or losing electrons, became electrically charged ions and the charge between them binds them together. In chemistry this is known as *ionic bonding.*

Richard Feynman said of quantum mechanics, *"You never understand it, you just get used to it."* Quantum vortex mechanics, with its simple models, is easy to understand. Not only can it help us comprehend quantum reality but it can help extend our horizons of understanding to new possibilities beyond the limited world view of materialism.

To break free of the limiting world view of particle materialism we need models in physics to replace the classical billiard ball model for the atom. The model of the quantum vortex of energy does just that. The model for energy in the vortex theory is a *line* of the movement of light either spinning in three dimensions to form a static quantum vortex or vibrating in two dimensions to form a transverse-wave quantum. These wave-kinetic particles of energy are treated not as vibrations in a ubiquitous electromagnetic field but as lines or strings of vibration.

A quantum of visible light is called a *photon* but this term is also used for wave-quanta outside the visible spectrum. The energy in each wave quantum, or photon of light, is proportional to the number of waves it contains. To appreciate this, imagine energy as a line or string folded into waves in a bundle of fixed size. The more waves in the bundle the greater would be the length of the string or line and therefore the more energy it would contain. For example, blue light is more energetic than red light as it has more waves in each photon.

This simple model explains why the energy in a quantum is equal to its

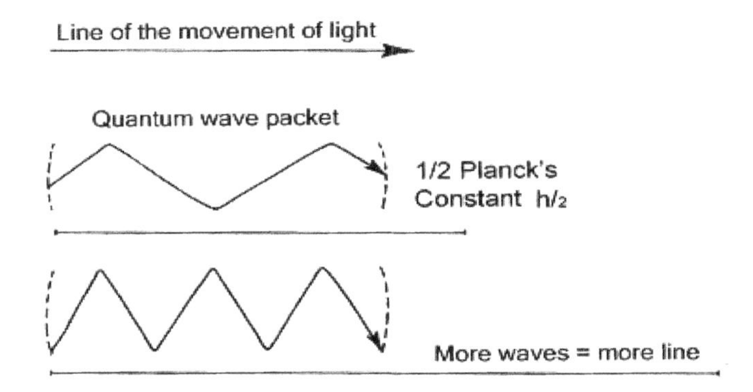

frequency times a *constant* called *Planck's constant*, named after Max Planck, the originator of quantum theory. In physics Planck's constant is denoted by the symbol 'h'. Planck's constant is applied to the wave-kinetic quantum which consists of two strings or lines of the movement of light, vibrating as transverse wave trains that are propagating at right angles to each other so if a wave-train bundle of energy were formed of a single line of the movement of light the energy it contains would be equal to half Planck's constant, denoted by '½h,' times its frequency of waves.

Single lines of the movement of light spin to form vortex particles of energy, such as electrons and protons. These are governed by half Planck's constant ½h, and are called after the Italian-American physicist Enrico Fermi, *fermions*. Photons of light and other radiant quanta, incorporating two lines of the movement of light are governed by Planck's constant 'h'. They are named after the Indian physicist Satyendra Nath Bose, *bosons.*

It is a mystery in quantum mechanics as to why quantum events occur as integer multiples of half Planck's constant, ½h or h/2. To appreciate why this is so and to understand another mystery, why in a wave-quantum, the propagating perpendicular transverse waves travel in opposite directions, we need to know something about antimatter.

CHAPTER 7
ANTIMATTER

Antimatter was first predicted in the 19th century. Writing to *Nature* in 1898, Arthur Shuster used the term *antimatter* for the first time and Karl Pearson proposed 'squirts' forming sources in matter and sinks in negative matter. Shuster hypothesized anti-atoms and antimatter solar systems and first proposed matter - antimatter annihilation.

The idea of antimatter surfaced again in 1931 when Paul Dirac predicted the existence of anti-matter in the math he was developing for quantum mechanics. It seemed like science fiction until later that year Carl Anderson at the California Institute of Technology spotted positively charged electrons emerge after bombarding lead with gamma rays.[1] Dirac had already called them *positrons.*

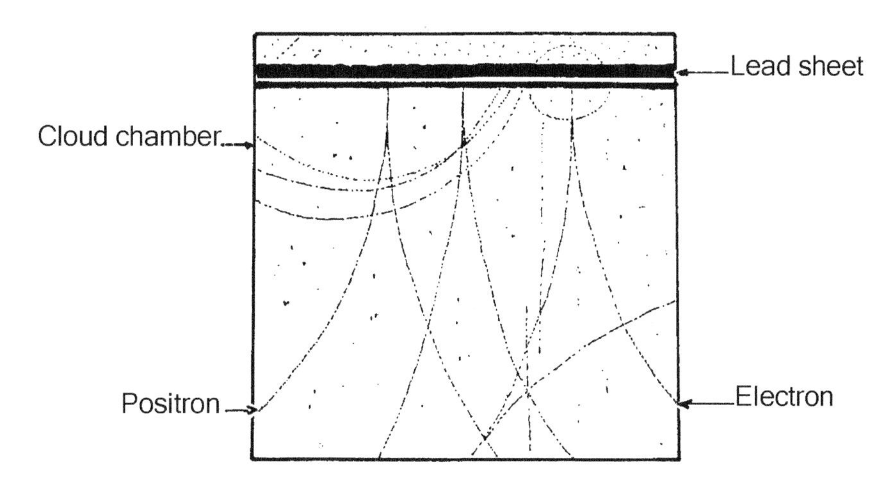

Anderson discovered that positrons were produced with electrons in pairs. He also realised only when gamma ray photons had energy exceeding the mass

1 Richards et al, Modern University Physics, Part 2, Addison Wesley 1973

energy equivalent of an electron and positron were the pair particles produced, so the production was obviously from a single gamma ray photon. This discovery confirmed that the photon consists of two strings or fields of energy each of which can be transformed into a particle of matter or antimatter.

In *This Way to the Universe: A Journey into Physics,*[2] Michael Dine relates how the prediction of antimatter was quickly followed by its discovery: *A crucial step was taken by Dirac. He made an inspired guess as to an equation that could describe the motion of a relativistic electron. This equation built in the known features of the electron (in particular a property called spin) from the beginning and accounted for many fine details of atomic structure. But the equation and its interpretation still posed serious challenges. The electron seemed to be unstable. Dirac, after some missteps, had another remarkable insight: He realized that the equation predicts the existence of antimatter – in this case a particle exactly like the electron in every way but with the opposite electric charge. With this rethinking of his equation, Dirac eliminated the problems...This particle was dubbed the positron, in deference to its positive electric charge. Dirac's theory predicted that an electron and a positron, if they meet can annihilate, producing other forms of energy (high-energy photons in this case).*

Dirac made his prediction in May 1931. The positron was actually discovered a few months later by Carl D. Anderson. Anderson was a brilliant experimentalist, and he was skeptical of theory and theorists. He made his discovery shortly after receiving his PhD, with an instrument he built to study cosmic rays, the energetic radiation (particles) from space that constantly strike the earth. Anderson was dismissive of the importance of Dirac's theory for his discovery, and clearly annoyed that Dirac's paper had appeared just a few months before: "Yes I knew about the Dirac theory...But I was not familiar in detail with Dirac's work. I was too busy operating this piece of equipment to have much time to read his papers...[Their] highly esoteric character was apparently not in tune with most of the scientific thinking of the day...The discovery of the positron was wholly accidental."

Anderson confirmed Shuster's speculation and Dirac's prediction that when matter encounters antimatter mass is annihilated. The product is generally two gamma ray photons that move off in opposite directions from the point of annihilation.

The gamma ray photons generated in matter-antimatter annihilation were used in an experiment conducted at Berkeley, in 1972, by John Clauser and

2 Dine M. This Way to the Universe: A Journey into Physics, Penguin Books, 2023

Stuart Freedman[3] who arranged for the annihilation photons to pass through polarizing filters on either side of the experiment. Behind these they placed photomultipliers to record them. When one photon was inverted they found that the other one flipped-over as well so that they were both plane polarized simultaneously.

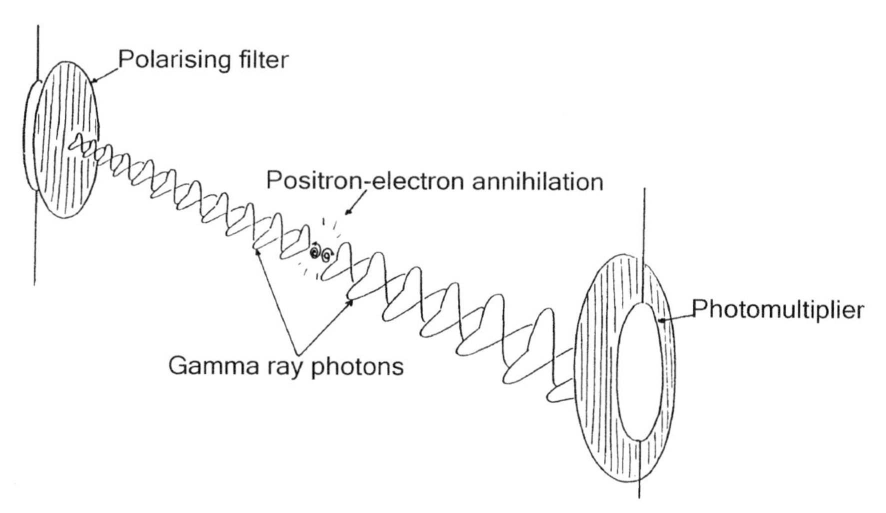

The Clauser–Freedman experiment appeared to confirm the model for the photon as a train of waves because of the way the photons seemed to be *entangled.* If the gamma ray photons entered the polarizing filters before they left the site of generation then, if the two gamma ray photons were entangled, the flip of one photon could cause the second photon to flip as well as it left the same point of generation but in the opposite direction. This entanglement would only be possible if the photons were trains of waves rather than vibrating particles.

Carl Anderson found that positron and electron pairs were produced only if the gamma ray photon encountered the nucleus of an atom. The quantum vortex account for this experiment would be that as each of the two wave trains of energy in the gamma ray photon drove through the nuclear vortices in an atom of lead they would be transformed into vortices in accordance with the first quantum law of motion. If the two subatomic vortices, appearing on the other side of the nucleus as an electron-positron pair, was each a quantum vortex of equal energy spinning in opposite direction that would account for their equal mass but opposite charge.

Anderson recorded the sum mass of the pair particles as being the equivalent

3 Clauser J. F. & Freedman S. J. Experimental test of local hidden-variable theories, Phys. Rev. 1972

to the energy of the original gamma ray photon, according to $E=mc^2$. He also noted the positron was drawn toward an electron by opposite charge attraction, which culminated in their mutual annihilation. In the quantum vortex account this matter-antimatter annihilation would be explained as two vortices of equal size but opposite direction of spin effectively unzipping one another.

The annihilation did not destroy the energy, it transformed mass back into the radiant form of energy. In the vortex theory this would be interpreted as the vortex form of the energy being transformed back into the wave form, in line with the second quantum law of motion. When the lines of the movement of light reverted from spin to their original wave form they radiated away in opposite direction as a pair of gamma ray photons.

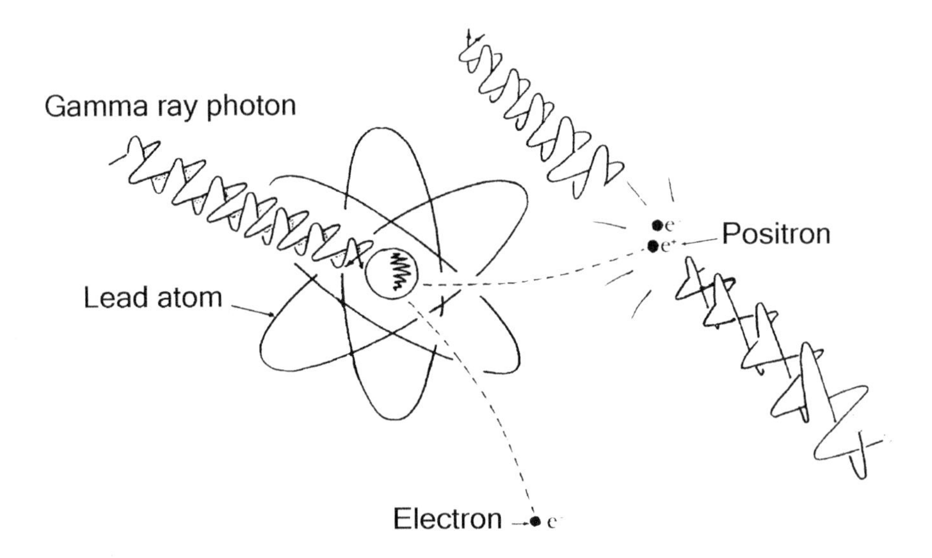

An interesting feature of the operation of the second quantum law of motion, in the annihilation of pair particles, is that the electron annihilated was not the electron newly produced but one already in existence. In the process, energy accounts appear to have been settled as the two photons produced by annihilation carried away the energy of the particle pair provided by the original gamma ray photon. This experiment appeared to confirm that while energy conservation laws apply in the overall process of production and decay of particles, conservation laws might not hold strictly in the steps between.

One line of the movement of light in the original gamma ray photon appeared to have formed an electron of matter whereas the other formed a positron of antimatter. That suggested half the quantum of energy was presumptive matter and the other half was presumptive antimatter. This experiment, dubbed *pair particle production and annihilation* supports the idea that every quantum of energy could consist of a presumptive particle of matter and a potential particle of antimatter.

Because antimatter does not exist in the normal world of matter it makes sense that one of the fields of energy in light would not react with matter. This has been confirmed in fluorescence and polarization where it has been shown that only one of the fields in the quantum reacts with matter and also in photography where only one field of energy in light reacts with the photographic plate. This suggests that each photon may be simultaneously in the world of matter and a world of antimatter. It could be that the wave-train quantum may be a closed loop in which energy travels in waves through the world of matter in one direction and then back through the world of anti-matter in the opposite direction. I call this *quantum loop symmetry.* Because each transverse wave train in the photon is a two-dimensional system of energy the quantum loop symmetry in the photon would be classified as *two-dimensional wave-quantum loop symmetry.* Quantum loop symmetry – depicted by the infinity sign – describes the endless cycle of energy between matter and unseen antimatter.

If the direction of time from past through the present and into the future were set up by the direction of waves in the photon then relative to us time would run backwards in the antimatter world. This fits with Paul Dirac's prediction of a reverse flow of time in antimatter.

If half of every quantum does not belong to the world of matter that could explain why in quantum mechanics events occur in integer multiples of half Planck's constant denoted by ½h. It would make sense that if only half a quantum is involved in the world of matter then only half a quantum would be involved in quantum mechanical events occurring here. This solution to the ½h mystery is suggestive of the existence of an antimatter Universe. Quantum loop symmetry infers if that were so it would be a precise mirror-image of the Universe of matter we occupy, not only in form but also in action. The quantum vortex theory provides a straightforward account for where the antimatter universe might be.

CHAPTER 8
THE MIRROR UNIVERSE

In the summer of 1969, as I sat by the stream in Crackington Haven, Cornwall or as I walked on the cliffs nearby, I pondered on the mysteries of the Universe. Staring at the stream it was obvious to me that the vortices and ripples, which appeared consistently in the same place, were just forms maintained by the constant flow of water. I realised the forms of matter might be similarly maintained by the constant flow of energy through subatomic vortices. The question in my mind was where did the energy spinning in these subatomic particles came from and to where did it go to. My concern was that the vortex theory must comply with the conservation law that energy is neither created nor destroyed

The water in the stream was part of a cycle of water between rain drops and the ocean. It struck me that like water in the *water cycle,* the energy passing through the vortices and waves that made me and everything I could see could be part of a universal cycle of vortex energy. I was working on my account for antimatter at the time and it struck me that the vortex energy in the Universe might circulate between matter and antimatter. I realised that vortex energy could be conserved if a parallel world of antimatter existed beyond the centre of every subatomic particle of matter.

Imagine energy spinning into the centre of an electron vortex as though it were a funnel. It then passes through a point of singularity at the centre, as though through a tunnel. Now visualise the energy spinning out the other side to form a *Siamese twin* positron vortex with equal mass but opposite charge. When I was working on my book[1] with Peter Hewitt, he pointed out that this image fitted Karl Pearson's idea of sinks and squirts.

For energy to form a uniform cycle it would have to circulate through every quantum vortex of matter in the Universe into an equal but opposite quantum

1 Ash D & Hewitt P The Vortex: Key to Future Science, Gateway Books, 1990

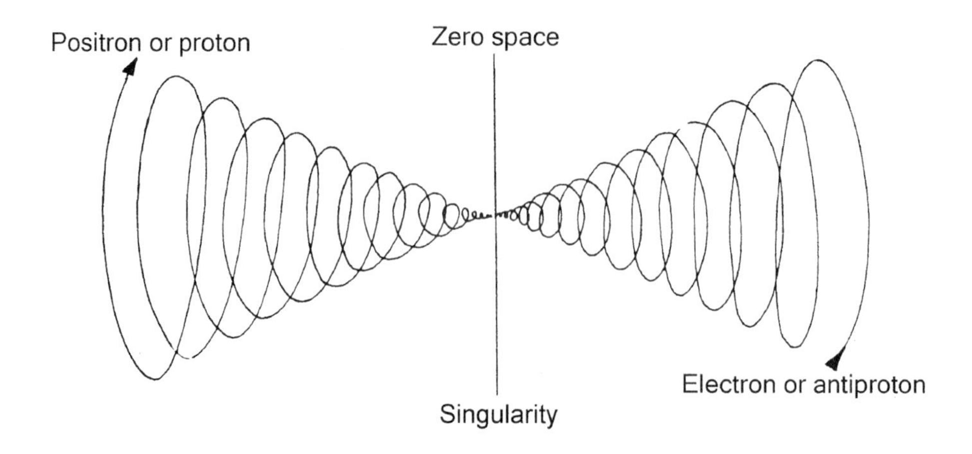

vortex of antimatter. That would mean beyond the centre of every subatomic particle of matter there would have to be an identical particle of antimatter. This implied a symmetry that every form and action in our world of matter would be faithfully replicated in a mirror world of antimatter. This vision of the Universe made sense of the fact that only half of every quantum of energy is effective in our world of matter. The other half could be acting in the world of antimatter. That would explain how, at a quantum level, all events may be duplicated in both halves of the Universe simultaneously.

That left me wondering if there might be an antimatter version of myself in an identical mirror half of the Universe pondering on the mysteries of the Universe by a stream of anti-matter water. The question I was left with was which was the real me and which was a mere reflection. The answer that came to me was just as I see a single world through a pair of eyes I could be simultaneously conscious in the two bodies. The simultaneous existence of every quantum in both halves of the Universe meant that every photon of light bringing me identical information in both worlds would strike my retinas in matter and antimatter simultaneously. Experiencing the exact mirror symmetry of all forms and actions in both halves of the Universe I would have a singular experience of reality.

The Universe, divided into mirror symmetrical halves, each acting as a reverse cycle of energy for the other, would be a depiction of quantum loop symmetry. As a three-dimensional system of spin energy the quantum vortex could be classified as a form of *three-dimensional spin quantum loop symmetry*. I envisioned an endless cycle of energy flowing in opposite directions in the Universe, either inwards or outwards, through looped pairs of quantum vortices, forming three-dimensional matter and anti-matter and the extensions of space between them. Again this is depicted by the infinity sign.

Quantum loop symmetry enables us to appreciate the Universe as being finite because it would not stretch on endlessly but also infinite as it would have no boundaries. The dual state of being finite and infinite is easy to understand. We know the Earth has a finite size but if we were to keep journeying on it we would never reach the ends of the Earth, we would simply arrive back where we started from. Just as we never leave one hemisphere of the Earth without entering the other so it would be impossible to leave the Universe of matter without entering the Universe of antimatter and vice versa. Each identical sphere of matter and antimatter would be separate from the other and both would have a finite size yet they would be infinite in their circle of connection. This model of a finite yet infinite Universe was proposed, in 1917, by Einstein as the *3-sphere model.*

Using the water cycle analogy for the universal cycle of vortex energy, I visualised a largest sphere of universal space as the equivalent of an ocean and the smallest spheres of space, at the *singularity* centres of quantum vortex particles of matter and antimatter, as equivalent to rain drops. These are where the centres of the vortex particles of matter and antimatter would connect and through which energy would flow, through smallest space, between the matter and the antimatter halves of the Universe. In my mind's eye I then imagined the concentric spheres of vortex energy extending out from matter and antimatter combining with the space extending from all the other stars and galaxies in both halves of the Universe. I visualised the matter and anti-matter meeting once again in the vast sphere of space connecting the two halves much as the meeting of two oceans. The vortex energy then flowed through the largest sphere of space between the matter and antimatter halves of the Universe.

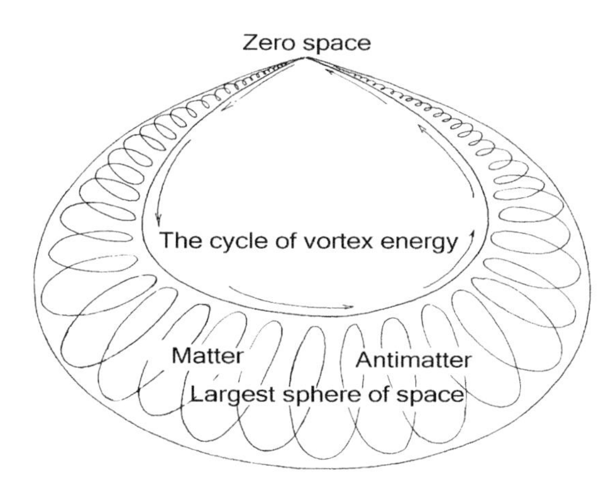

In the water cycle water circulates between the large and the small, between oceans and rain drops. I imagined that like this a universal cycle of vortex energy circulated through matter and antimatter. Between the extremes of the miniscule in an atom and magnitude beyond outermost galaxies energy flowed between the smallest and largest spheres of space.

I found it difficult to comprehend the loop of a photon enabling it to be in the realms of matter and antimatter simultaneously just as I found it difficult to imagine how the two halves of the Universe could share a common surface and common centres and yet remain entirely separate. I realised the problem I had was that I was limited in my comprehension of the Universe by my mundane experience of it. I had to accept that the Universe is not limited by my perception of it.

After the summer recess I was looking out of a window at college and I saw more examples of symmetry. I realised just as the cars on Campden hill couldn't drive without a road so motion could not exist without space to move in. It dawned on me that each vortex and wave train of energy might be effective as a car to every other vortex as a road.

Imagine each vortex of energy acting as a system of activity *relative* to every other vortex passively providing for it as a void of space. Each and every quantum vortex, as a form of motion, would depend on the three-dimensional extension of every other vortex of energy in existence for their ability to move. This existential co-dependence between vortices of energy could be viewed as an expression of a *universal law of love*; that everything in the Universe depends on everything else for its existence. As photons of light also depend upon the extensions of vortex energy to provide space through which they can move everything in the Universe, both matter light, would be in an ocean of love.

In the spring of 1970 frog spawn inspired me to imagine space as foam. I imagined the black frog embryos at the centre of the eggs as the massive centres of the vortex particles and the frogspawn foam as *space foam* extending from them. Each frog egg was in its own bubble of space in the foam. As a three-dimensional extension it contributing to the collective foam of all the other eggs and as I stared at the tadpoles growing in their eggs I realised they were growing simultaneously in the three dimensions of space. This provided me with a model for a *fourth dimension* in space in which things shrink or grow.

In *Alice's Adventures in Wonderland* the fourth dimension was epitomised in the story of Alice chasing time. Down a rabbit hole she experienced size shrinking to enter a looking glass world. That fitted my thinking of shrinking space leading into a mirror symmetrical world of antimatter. The twins *Tweedle Dum* and *Tweedle Dee* agreeing to do battle depicted particles of matter and antimatter annihilating each other, so I decided to call the dimension of bigness and smallness after Alice, the *Alician dimension*. Later I also called it the *fractal dimension* as fractals represent repeating identical patterns in different dimensions of size. In linear movement things move up or down, forwards or backwards, left or right in one of the three dimensions of space. In the *Alician* or *fractal* dimension, things move by growing or shrinking in all three dimensions of space simultaneously.

The *Alician Dimension* is essential in the theory of the quantum vortex because the subatomic vortex of energy is a system of concentric spheres of energy continually growing or shrinking (according to their sign of charge). The fractal or Alician dimension of contraction and expansion as the fourth dimension was important for me in understanding time.

It struck me that the vast distances of space and aeons of time might be temporal illusions set up by the dimension of size. Our experience of size would be relative to our position in the spectrum of size in space. Maybe intergalactic space only appears vast to us as we are little compared to a galaxy. As it might take forever for a bacterium to cross the road because it is tiny compared to us, whereas we can nip across a road in a moment, so galaxies might not take long, in galactic time, to interact with neighbouring galaxies. This imagining helped me to comprehend how a minute quantum of wave-form energy could be looped between matter and anti-matter over the vastness of space. As a quantum wave-train of vibrating energy is not an extension in three dimensions it would not be constrained to the Alician 'size' dimension so it would not be bound by size.

Thinking about time I wondered if perhaps the uni-directional progress of light through space might set up the uni-directional progress of time from past through the present and into the future. Einstein defined time as the fourth dimension and I saw a link between time and the size of space acting as a fourth dimension. Einstein suggested time dilates with acceleration. I was aware that space contracts as energy accelerates into the centre of the vortex of energy. I also knew that time dilates in the shrinking realms of contracting space. I came to this conclusion after chancing through the students union one day when a television was showing a film of cats and insects slowed down. In dilated time the cats moved with the fluidity of panthers and the insects flew like birds. It struck me their movements only appeared fast and jerky to me because I was bigger than them. If I were the size of a mouse a cat might come at me like a panther and butterflies would wing as birds. I realised that my experience of time was relative to my position in the size spectrum of space. I imagined if I were a space giant big enough to view the Earth as a football, I might use its motions as my clock. As a space giant I could use the spin of the Earth to count out my seconds, the orbit of the moon to be my minutes and the annual circuit of the Earth round the sun to mark out my hours. My years would be centuries to people on Earth who would be to me as bacteria on a football. My lifespan would be millennia to them. If I was as big as the Universe a billion years to humans on Earth might be as a day to me. On that scale the Universe could have been around for about a fortnight and the appearance of the Earth and evolution of life upon it might have happened in a matter of days. Realising that my experience of time is relative to my size I reconciled my science with the days of creation in the Bible.

I delighted in Einstein's special theory of relativity. His special theory was compatible with my quantum vortex theory and was essential to my understanding of matter, space and time. But not so his general theory of relativity. I found it complex and difficult to understand and it did not fit with the account for gravity that was beginning to unfold before me in the progress of the quantum vortex theory.

CHAPTER 9
GRAVITY

After publication of his first theory of relativity in 1905, Albert Einstein published a second theory in 1915 in which he proposed gravity was caused by a distortion of space and time by mass. Einstein contended that together space and time formed a four-dimensional *space-time continuum*, which could be bent by mass resulting in a curvature of space and a dilation of time. To model his idea Einstein used a sheet of rubber to represent space-time and heavy balls to act as the massive bodies of matter that distort it.

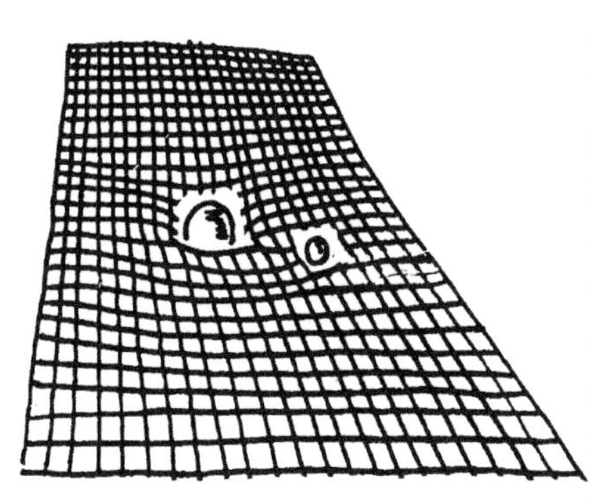

In the rubber sheet model, the ball stretches the rubber and creates a dip. The stretching of the sheet represents the stretching of time and the dip depicts the curvature of space. As another ball enters the dip it rolls in the curve toward the first ball. Einstein suggested that the cause of gravity was an acceleration of the second body of matter toward the first in the contoured space-time continuum. He argued that this effect occurred not because the object entered a gravitational force field but because it entered a region in which space-time was distorted by a massive body of matter. The result of entering a region where time slowed down would be that the object would appear to accelerate.

In his general theory of relativity, Einstein predicted that light would appear to be deflected in a curved path by the space-time distortion. This effect could be observed during a solar eclipse when the space-time distortion around the sun would result in an apparent deflection of starlight. He suggested this could be measured by comparing the position of the stars around the sun during an eclipse against their normal position in the night sky.

In 1919, The Royal Society launched two expeditions on separate continents to test Einstein's prediction. One was to Sobral in Brazil. The other was on the other side of the Atlantic, in Principe, on the west coast of Africa. On both expeditions photographs were taken of stars around the sun during the solar eclipse. The same stars were then photographed in the night sky.

Back at the Royal Society in London the photographs of the sky at night were compared with the photographs of the sky during the eclipse. The scientists saw that the positions of the stars close to the sun appeared to have changed as predicted. Obviously the stars had not changed their positions so it was accepted that the effect must have been deflections of starlight during the eclipse caused by the warping of space-time by the mass of the sun, just as Einstein had predicted.

On November 7th 1919, The Times of London published an article headlined, *"Revolution in Science, New Theory of the Universe, Newtonian Ideas Overthrown,"* which sparked a media frenzy that caused Albert Einstein to became the most celebrated scientist in the world.

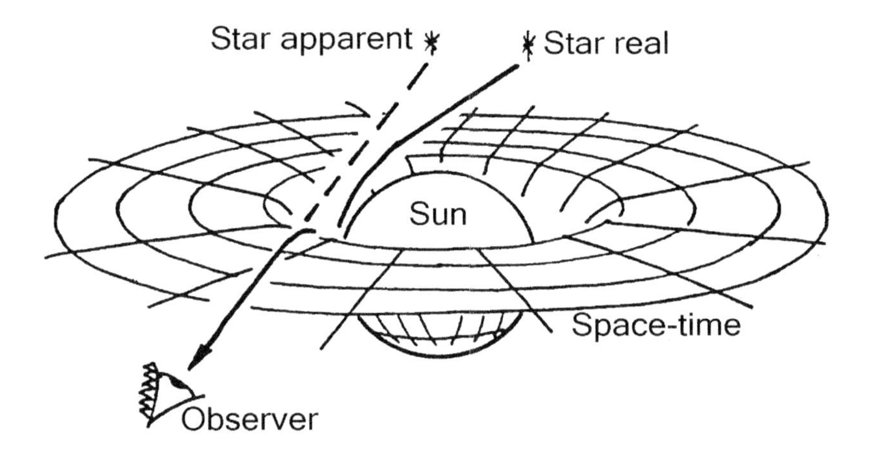

I disagreed with Einstein's theory. I didn't believe the space-time around the sun was distorted by its mass. In my theory the space around the sun is curved not because it is distorted by the mass of the sun but because it is an extension of the shape of the sun.

According to the vortex theory the space around the sun is curved because space in the immediate vicinity of the Sun would be a collective of the shells of space extending from every quantum vortex in the sun. As such the sun's space would be an extension of the spherical shape of the Sun. The light from stars following the curvature of space would then be deflected round the sun.

I envisaged starlight following the curved space path round the sun much as a motor car follows the curvature of a road around a roundabout.

In the quantum vortex theory I predicted the same outcome as Einstein predicted in his general theory of relativity but it was much simpler. My account for space curvature was an integral part of my approach to quantum theory whereas Einstein's theory was difficult to integrate into the quantum theory he had established.

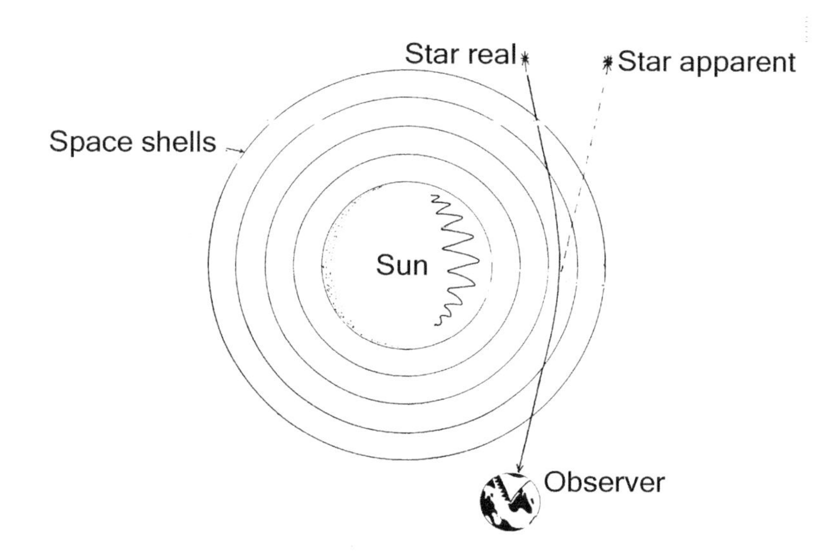

If Einstein declared to a reporter that the summation of his relativity theory was a connection between matter and space-time why didn't he consider that space-time might be discontinuous. It is obvious, if matter is particulate the space-time connected to it must be particulate too. In my theory I felt I had trumped Einstein by presenting a quantum theory for space so I was confident that a better theory for gravity than his might emerge from it.

Like a detective I was looking for clues. I knew from the similarity of the equations between electric charge and gravity it was likely that the force of gravity was a vortex interaction. Another clue was that electric charge is made up of individual charges contributed by particles of matter. Gravity, likewise, was made up of components of gravity from individual particles of matter.

This fact alone enabled me to draw a straight line between the vortex of energy and gravity. Yet another clue lay in the fact that just like electric charge, gravity obeyed the inverse square law in that the intensity of its action diminishes by the square of the distance from the centre of action. I recognised that was a signature characteristic of the spherical quantum vortex.

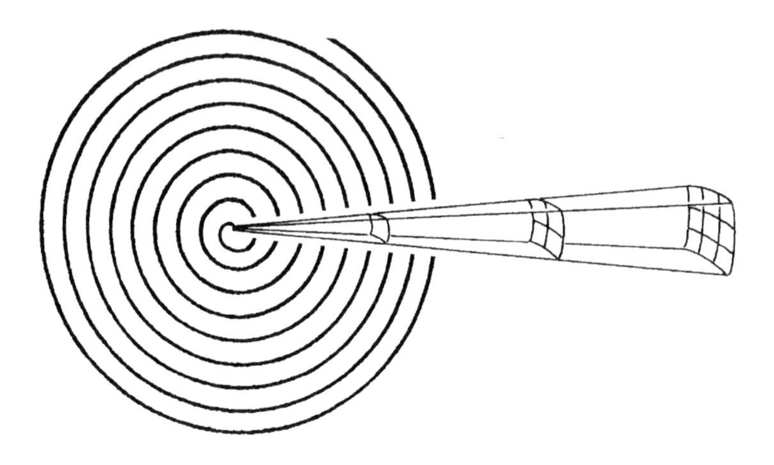

The clues that gravity was caused by a vortex interaction were coming in thick and fast. The forces of electric charge, magnetism and gravity all caused an acceleration and I knew that acceleration into the vortex centre was a feature of the quantum vortex.

I found a book in the college library that suggested gravity may be linked to the acceleration of energy into the centre of the vortex. In the book I read about Einstein's *principle of equivalence of forces* proposed in his general theory of relativity. It came from one of his thought experiments in which he pointed out that the occupants of a sealed rocket ship would experience a fall toward the rear of the rocket as it accelerates away from the Earth but if the rocket were to fall back to Earth its occupants would experience weightlessness. He argued that the occupants of the rocket ship would not have been able to distinguish between the force of gravity and the acceleration of the ship and concluded that we could not be certain whether gravity was the result of an acceleration or the action of a force field.

Since the media storm in 1919 it has been generally accepted that Einstein toppled Newton with the general theory of relativity as he appeared to have dismissed Newton's force field view of gravity but it struck me Einstein's principle of equivalence of forces would not allow that. His principle did not allow for an acceleration theory for gravity to be truer than a force field theory if they were points of view because, as he himself emphasised, experiments cannot distinguish between points of view. The principle suggested that Newton and Einstein could both be correct in their theories for gravity if they were shown to be points of view.

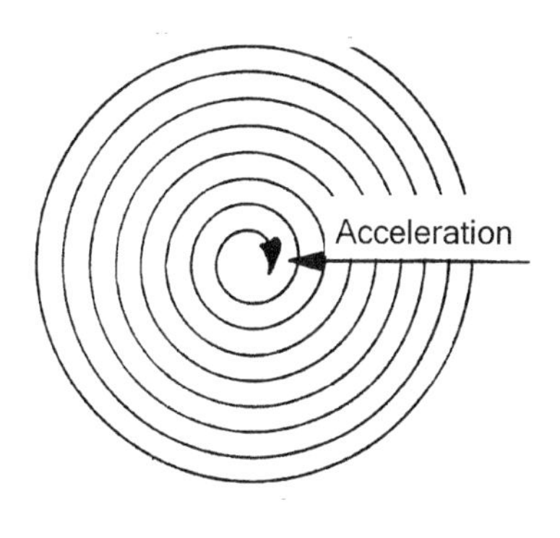

I realised the quantum vortex theory configured with both Newton and Einstein's approaches to gravity in 1972. After a recording session, in a studio just off Holland Park Avenue, I was transfixed by reels fast winding on the recording decks. As I watched the spools accelerate, when the tape wound off them in a spiral toward the centres, I saw vividly how spiral motion toward the vortex centre would result in an acceleration down its radius. Walking distance from the library where I first encountered Einstein's principle I was in a studio that revealed just how well it could work in the quantum vortex theory.

When spheres are placed in contact they touch at the point of a radius. As a vortex of energy is effective as an acceleration down the radius toward its centre, from the perspective of one vortex relative to another they would appear to be an acceleration. This represented Einstein's point of view. However, from the perspective of an observer looking down on a pair of overlapping vortices of energy they would appear as interacting force fields. This represented Newton's point of view. As no one could determine experimentally whether one point of view was more correct than the other a comprehensive theory for gravity had to incorporate both Einstein's and Newton's points of view.

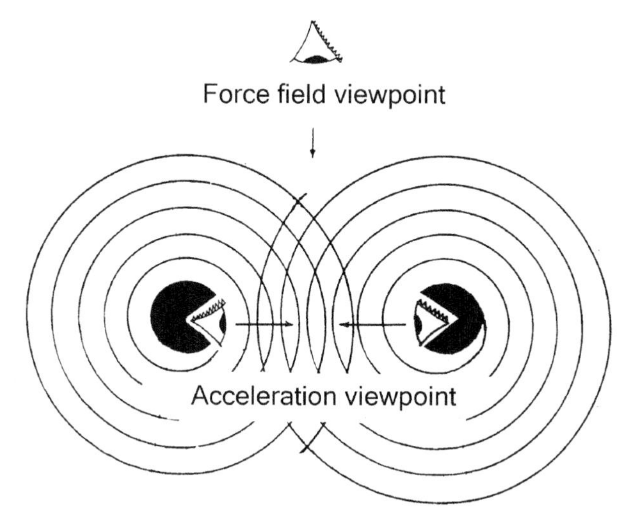

The Royal Society observation of the apparent deflection of starlight around the sun was thought to confirm Einstein's theory. However, the 1919 observation could equally confirm the view that concentric spheres of vortex energy extending from the sun caused the starlight to be deflected. This view would support the model of a force field if the concentric spheres of vortex energy extending from the sun were to transmit the sun's gravity into space. I was developing a theory for gravity that showed it might be caused by vortex interactions, the effect of which could be transmitted into infinity by the sun's extending field of space.

I was working on the matter-antimatter model for the Universe when it struck me there might be a connection between the force of gravity and antimatter. This was because the force of gravity came from the centre of matter and in my model, antimatter occurred beyond the centre of every subatomic particle of matter in existence. I realised that my idea of a vortex particle of antimatter, beyond the centre of every vortex particle of matter, each with an opposite charge, was supported by Lord Rutherford's *Law of Conservation of Electric Charge,* which declared that for every charge in the Universe there is an opposite charge.

Particles of matter and antimatter are opposite in charge and between opposite charges there is an *accelerating force of attraction.* I proposed that this could account for the accelerating force of attraction we experience as gravity. If there is antimatter beyond the centre of matter there could be a vortex interaction acting through the collective centres of subatomic particles causing them to *centralise* on each other. This is the hallmark feature of gravity.

The centralising pull of gravity, acting through the centre of matter, would be transmitted through the entire vortex extending into infinity. If you pull a person by a finger the person will follow the pull because the finger is an integral part of that person. Likewise, the spheres of space extending from the sun would transmit the sun's gravity into infinity because they are an extension of the sun. They would extend into infinity the collective of gravity contributed by every vortex of energy in the sun. I realised that the pull between matter and antimatter could not be between the *connected Siamese twin particles* because they belong to the same system of vortex energy. The gravitational force of attraction could only be between a body in matter and body in antimatter, which are not connected. For example, there would not be a gravitational vortex interaction between the particles in my body in matter and my body in antimatter because they belong to the same quantum loop so they could not move relative to one another. However, my body of matter could experience a pull due to a force of attraction coming up from all the other vortex particles in the anti-matter Earth acting through the centre of the matter of the Earth. This is the evident action of gravity. Every single body on the Earth is attracted by a force of attraction acting from the centre of gravity of the Earth pulling us all and everything else toward the Earth's centre. The vortex account for the Earth's gravity which keeps us all attached to it contradicts Einstein's opinion that the Earth's gravity is caused by the mass of the Earth distorting space-time. In the quantum vortex theory each of us is pulled toward the centre of the Earth by a force of attraction that is coming up from the antimatter Earth through the body of our planet of matter. This is then transmitted through the Earth's quantum field of space into infinity. In the quantum vortex theory gravity is treated as an action of space. You can see from the table below; gravity and space have a lot in common.

	Gravity		Space
1.	*Extends from matter.*	1.	*Extension of matter*
2	*Acts from the centre of each particle of matter*	2.	*Extends from the centre of each particle of matter*
3.	*Unlimited range in 3D*	3.	*Unlimited extension in 3D*
4.	*Obeys Inverse Square Law*	4.	*Obeys Inverse Square Law*
5.	*Curves the path of light*	5.	*Curves the path of light*
6.	*Universal gravity is the sum of the gravity acting from every particle of matter*	6.	*Universal space is the sum of space extending from every particle of matter*

Einstein invented a story for gravity. Mine is a different story. Both stories account for the deflecting effect that massive bodies, like the sun, have on starlight. Einstein's story is built on the arbitrary element that mass distorts space-time. My theory it simpler and easier to understand than Einstein's *general theory of relativity* and I can invoke *Ockham's razor* on the basis that I use the same theory of vortex interactions to account for gravity and electric charge and magnetism.

I place gravity in continuum with space and mass, electric charge and magnetism as aspects of the vortex of energy. Because the quantum vortex is formed of movement at the speed of light, gravity, space and mass are relative to this invariable speed. Because the quantum vortex is curved, space is curved. Because energy in the quantum vortex extends into infinity, bodies of matter can act on each other over seemingly infinite distances. As quantum vortices are intrinsically dynamic arbitrary elements are not needed to account for their interactions. According to the quantum vortex theory vortices of energy do not interact due to the mediation of arbitrary force fields existing apart from them, they are the force fields. No arbitrary assumptions are layered on the single axiom that subatomic particles are vortices of energy. In the theory of the quantum vortex the forces and properties of matter and antimatter are accounted for in a single, unified field of logic.

CHAPTER 10
BLACK HOLES

In 1972 I realised I could use the vortex theory for gravity to explain black holes. My premise for black holes came from reasoning that if quantum vortices of energy can overlap and occupy each other's space then it is conceivable they could all coincide on the space of a single subatomic particle. In my theory the entire Universe could collapse, due to gravity, onto a singularity point less than the size of an atom. The electric force of repulsion between protons works against gravity and prevents this collapse from happening. Protons packed in the nucleus of an atom do not converge into singularity because the force of electric charge pushing them apart is stronger than the force of gravity pulling them together. However, in the demise of stars, when the loss of hydrogen fuel reduces the kinetic energy of electrons, the electrons can fall into protons to form neutrons. Then with the electric force cancelled out, the force of gravity can take over. That is when the collapse of matter into singularity becomes possible. In this process small stars can become dwarf stars, more massive stars can collapse into neutron stars and the most massive dying stars can become black holes. Representing the greatest density of matter, black holes are the greatest concentration of space in the Universe.

When stars in matter collapsed toward singularity the same would be happening to their antimatter counterparts and as they converged annihilation could occur between them. This led me to conclude that the energy of annihilation could be released by the action of gravity and the release would occur through the singularity points in the core centres of black holes. *Quasars* - quasi-stellar phenomena - had been recently discovered and I thought they might be evidence of this.[1]

Quasars are amongst the most energetic things in the Universe. Releasing nearly two hundred times the energy of stars, their energy profile fits matter-antimatter annihilation. (The energy release of nuclear fusion in stars

1 Greenstein J & Schmidt M The Quasi-Stellar Radio Sources. Astrophysical Journal 140, 1964

comes from the burn of 0.7% of the mass of a proton. In matter-antimatter annihilation 100% of the mass of a proton is consumed which releases around two hundred times the energy of nuclear fusion.) Quasars have been discovered at the centres of galaxies where super-massive black holes are found. This is where I expected them to be if their energy came from matter-antimatter annihilation through the smallest realms of space. But quasars have also been discovered at the furthermost reaches of the Universe, which suggested to me that annihilation was also occurring across the largest sphere of space. This is what I expected because that was where my theory suggested matter and antimatter meet and annihilation would again occur. The discovery of gamma ray bursts in outer space in the late 1960's added strength to this premise because gamma rays are the hallmark of matter-antimatter annihilation.

Gamma ray bursts are now known to originate from outside our galaxy and to be a thousand times more luminous than quasars and one hundred quadrillion times more luminous than the sun. More recently they have been confirmed to be associated with the transition of massive stars into black holes and subsequently periodic gamma ray bursts were discovered to occur from black holes. Now they are most commonly associated with the core collapse of very massive stars into black holes.[2]

Gamma ray bursts provide evidence that antimatter exists beyond the core centres of stars as predicted in the quantum vortex theory and that gamma rays are released from black holes where mass is most centralized. It is abundantly clear that matter-antimatter annihilation occurs where gravity is strongest. There can be no doubt that antimatter exists beyond the centre of matter. But why do the gamma ray bursts occur periodically? In the vortex cosmology there is a simple answer. Black holes could be behaving somewhat like *Old Faithful* in Yellowstone Park. If the theory of the quantum vortex is correct they should release gamma rays in the manner of a *geyser*. Let me explain. In the quantum vortex theory gamma rays would be produced continuously by the annihilation of matter and antimatter through the centre of a black hole. In line with the first quantum law of motion the black hole would act as a gargantuan proton vortex holding the gamma ray photons in complete capture. Initially they would not be free to escape but with the continual build-up of annihilation energy there should come a point when the amount of energy in capture would exceed the ability of the gravity of the black hole to hold it. Then, in line with the second quantum law of motion, the annihilation energy could escape from capture, as a burst of gamma radiation, after

2 Fishman C. J. and Meegan, C. A. Gamma-Ray Bursts, Annual Review of Astronomy and Astrophysics 33 1995

which the cycle of capture would begin all over again. I called this process the *gravity geyser.* In attempts to predict the periodicity of gamma ray bursts from supermassive blackholes in galactic centres, account should be taken of their random devour of stars as that would increase their mass, and so their gravity, which in turn could prolong the gravity geyser intervals.

Recent discoveries of high-speed streams of electrons as well as gamma ray emissions from the centres of galaxies support the idea that annihilation of matter and antimatter is going on inside the black holes located within them. William Tiller of Stanford University and William Bonner of London University lend support to the quantum vortex cosmology by relating this to gravity and hidden properties of a mirror world.[3]

The evidence of antimatter beyond the core of black holes is accumulating. It is provided by bursts of gamma rays, high energy electrons and X-rays now recorded regularly as astronomical events. Increasingly this is supported in the scientific journals In a *Scientific American* article, for example, T*he Brightest Explosions in the Universe*,[4] account was given of annihilation between matter and antimatter through the centres of black holes.

A theoretical model in physics is needed to account for antimatter beyond the centre of matter. The theory of the quantum vortex does that. The existence antimatter beyond the centre of matter is essential to the working of the theory because it completes the universal cycle of vortex energy necessary for the theory to comply with the first law of thermodynamics that the total energy of a system remains constant.

Unfortunately cosmologists have dispensed with antimatter. There is none left in their ledger to account for the gamma ray bursts. Sadly leading cosmologists firmly believe that at the big bang there was slightly more matter than antimatter. In this arbitrary assumption the antimatter immediately annihilated with most of the matter and the remnant matter is what we see as the Universe.[5] That is a feeble argument in my opinion!

In 1995 I published the vortex cosmology in which I predicted there is equal antimatter to matter in the Universe.[6] Predictions of an antimatter half of the Universe go back to Paul Dirac and the theory of a *Trouser Leg Universe* from Tel Aviv University in the 1970s. These proposals were dismissed for lack of evidence of gamma rays, which is the signature of matter- antimatter

3 Tiller W. Science and Human Transformation, Pavior, 1997

4 Gehrels et al, The Brightest Explosions in the Universe Scientific American Dec 2002

5 Hawking Stephen, Black Holes and Baby Universes, Bantam 1993

6 Ash David, The New Science of the Spirit, The College of Psychic Studies, 1995

annihilation. Despite the fact there is now an abundance of gamma ray evidence there is still resistance to the idea of an unseen underworld of anti-matter beyond matter. The only possible reason for this can be the inertia of consensus opinion. As Thomas Kuhn said, *"Truth has as much to do with the consensus of scientists as to the outcome of experiments"*.[7]

On January 15th 1975, I gave a lecture to young members at the Royal Institution in London. Toward the end of the lecture I explained that because of the unseen world of antimatter, galaxies of matter could accelerate toward the largest sphere of space due to the electric attraction acting between them and antimatter beyond the greatest extremity of space. I also predicted that eventually they would meet and annihilate and then I proposed that quasars were evidence of this. In that lecture I also predicted that due to the accelerating attraction between matter and antimatter, in the expansion of the Universe the furthermost galaxies would be accelerating away from us faster than the more proximate ones. I published this prediction two decades later in my 1995 book, *The New Science of the Spirit*[6] and I also suggested that because the acceleration of the galaxies we observe would be toward galaxies of antimatter it would lead to the annihilation of its mass rather than an indefinite expansion of the Universe. I argued that the existence of quasars in the outermost reaches of the Universe was evidence of this.

The presence of quasars at either end of the size spectrum of space, occurring in the largest as well as the smallest realms of space, adds credence to the vortex cosmology and increases the credibility of the theory of the quantum vortex. Occurring in the regions of densest gravitational contraction and greatest universal expansion they fit the predictions of where matter meets antimatter. However, annihilation of the vortex form of energy would lead not only to the end of matter but also to the end of space and the forces of electric charge, magnetism and gravity associated with matter. If this occurs at both ends of the spectrum of size then there could be a point. However, I do not think that is what actually occurs.

Imagine a demise scenario for the Universe. First exhausted stars collapse in on themselves to form neutron stars, dwarf stars or black holes. Then super-massive black holes at the centres of galaxies engulf them along with millions more exhausted stars in their gravitational maelstroms. Our modest galaxy has a moderate super-massive black hole at its centre called *Sagittarius A star.* It houses seventeen million exhausted stars in a sphere that would occupy the space of our sun out to the orbit of Mercury. Many galactic cores are much more massive and engulf hundreds of millions of dead stars. These

7 Kuhn T. The Structure of Scientific Revolutions, University of Chicago Press, 1962.

black holes appear to be massive graves where stellar carcasses are buried. This outlook appears bleak. All that's left would be darkness.

However, I don't believe the Universe ends in either light or dark. In the vortex theory a different picture is painted. In the vortex cosmology when a black hole forms in matter, an antimatter black hole also forms in antimatter. As stars of matter in the black hole are centralized into singularity, they would be converging on stars of antimatter in the mirror black hole, which would also be converging into its zero-space point so annihilation between matter and anti-matter would take place. With annihilation, the mass of both black holes would decrease and their gravity would diminish so their survival would depend on assimilating more dead and dying stars in matter and in antimatter. Due to annihilation, the energy released by both black holes would be continually increasing until it overcomes the grip of gravity then it could escape and radiate away. Stephen Hawking determined that black holes leak energy which comes from matter-antimatter annihilation. In my view if annihilation destroys space, time, mass and gravity and if black holes transform into quasars, then black holes may be only temporary astronomical formations, just steps in the transition of stars into quasars; of vortex energy into wave energy; of mass into light.

If black holes do undergo transition into quasars then according to the vortex cosmology many black holes could have a quasar associated with them. In fact, new quasars are being discovered with black holes at their core that are emitting super-fast streams of electrons and gamma ray bursts perpendicular to the black hole centre. The bursts appear suddenly and then die away. The energy and behaviour of these emissions may be perplexing to astrophysicists,[3] but they fit with the cosmology of the quantum vortex. The periodicity of these emissions need not be perplexing if they are seen to come from a tussle between the gravity causing particles to conglomerate and the energy of annihilation tending to blow them apart. This is how a gravity geyser would be expected to perform.

High energy electron streams and gamma ray bursts appear to confirm that there is matter-antimatter annihilation occurring through the intense gravitational cores of black holes. I like to think of these cosmic formations not as astral graves but more like stellar pupa in which astral caterpillars morph into quasar butterflies.

In the vortex theory gravitational energy is seen as a product of annihilation between matter and antimatter. That fits with Einstein's belief that gravitational energy is derived from the destruction of mass. When a body falls due to gravity it releases energy on impact. This energy would come from the destruction of mass because a moving body has a greater mass than a body at rest.

As a body falls in the world of matter a mirror-image body of antimatter would also be falling so both would make their impact and release energy simultaneously. There would be an equal minute annihilation of mass in the matter and antimatter as both bodies come to rest. At rest they would both be closer to their ultimate annihilation.

The fall of every bit of matter on Earth is another step in the inexorable progress toward the ultimate annihilation of mass in the Universe. As you pick up an object that has fallen you use up energy from the sun that was acting to keep the particles in the sun from falling together due to gravity. Eventually, when the sun's reserves of fuel are exhausted there will be nothing left to keep you alive let alone lift things! The inertia of gravity will take over and all the matter within our local star will collapse toward antimatter and annihilation.

Even though the sun is too small to form a black hole and will first become a *red giant* before eventually settling down as a *white dwarf,* ultimately it will be swallowed by the super-massive black hole at the centre of our galaxy. That will occur as the entire galaxy consumes its reserves of nuclear fuel and collapses under the awesome influence of gravity.

Annihilation may be an ultimate transformation of matter into light but in the quantum vortex theory the demise of space and gravity would come with the annihilation of mass. Before all the mass of the universe is annihilated, before every vortex particle in matter and antimatter can unzip, gravity would be declining and the increasing energy of annihilation would be compressed into collapsing space. It would be forced to race in an increasingly tight spiral space path with less gravity to grip it. This would set a powder keg for explosion. In the conglomeration of collapsing galaxies as their mass and energy is compressed toward singularity into smaller space, taking into account the decline of gravity, a big bang explosion would be inevitable.

The death of the Universe could lead to its rebirth. The scene would be set for a new cycle in the Universe and a new phase of expansion to begin. The big bang might not be a one-off event in the history of the Universe. If a big bang happened in the past maybe another big bang will occur in the future. Maybe the big bang was not the beginning of the Universe. Maybe it signaled a rebirth of the Universe.

Sages in India speak of the comings and goings of the Universe as the in-breath and out-breath of Brahma. Brahma, Vishnu and Shiva represent the birth, life and death of the Universe but Shiva, the god of destruction, is also the god of rebirth and recreation. There is no finality in the Hindu cosmology, only an endless cycle from birth through life to death and then rebirth again. We see this happening around us where apparent beginnings and ends turn out to be part of a cycle of renewal and regeneration. This may apply to the Universe as a whole. If energy is neither created nor destroyed maybe the Universe was never created nor will it ever be destroyed. All we know for certain is that the Universe of energy exists. All we ever observe are its unending transformations.

CHAPTER 11
DARK ENERGY

Prior to 1990 it was assumed from the big bang theory that the Universe was expanding at a uniform rate. But in 1990 Saul Perlmutter, professor of Physics at Berkeley University, decided to test that assumption. He pulled together a team in the USA and the UK to use redundant telescopes in a search for supernova explosions in distant galaxies. After half a decade they had spotted a sufficient number of the rare events to draw a remarkable conclusion. In the December 1997 publication[1] of their discovery Perlmutter and his team disclosed that the expansion of the Universe is accelerating.

To quote the Astronomer Royal, Martin Rees: *"An acceleration in the cosmic expansion implies something remarkable and unexpected about space itself: there must be an extra force that causes a 'cosmic repulsion' even in a vacuum. This force would be indiscernible in the Solar System; nor would it have any effect within our galaxy; but it would overwhelm gravity in the still more rarefied environment of intergalactic space. Despite the gravitational pull of the dark matter (which acting alone would cause a gradual deceleration), the expansion could then actually be speeding up."*[2]

In *A Brief History of Time*,[3] Stephen Hawking wrote: *"A theory is a good theory if it satisfies two requirements: It must accurately describe a large class of observations on the basis of a model that contains only a few arbitrary elements and it must make definite predictions about the results of future observations."*

In my 1995 book[4] I published a *Cosmology of the Vortex* in which I predicted that the expansion of the Universe is accelerating. That gave me a two year

1 Perlmutter S. et al Discovery of Supernova Explosions... Lawrence Berkeley National Laboratory, Dec 16, 1997 and, Discovery of a supernova explosion at half age of the Universe. Nature, 391, 51-54, Jan 01, 1998

2 Rees M. Just Six Numbers, Weidenfield & Nicolson 1999

3 Hawking S. A Brief History of Time, Bantam Press, 1988

4 Ash D, The New Science of the Spirit, The College of Psychic Studies, 1995

priority over Perlmutter's publication. In 2011 he received the Nobel Prize for his achievement.

Perlmutter called his discovery *dark energy* because he said he was in the dark about the cause of it. I was not in the dark about the cause of the accelerating expansion of the Universe. I knew it was the polar opposite of gravity on a cosmic scale.

In my opinion the account the accelerating expansion of the Universe was more important than predicting it before it was discovered. There is a precedent for this. Albert Einstein was awarded a Nobel Prize not for his predictions in the general theory of relativity but for the way he used Max Planck's quantum idea to account for the *photoelectric effect*. That was how he established quantum theory.

Nonetheless *The Quantum Vortex* qualifies as a good theory because it was successful in predicting the result of a future observation as well as accurately describing a large class of observations on the basis of a model that contains few arbitrary elements. In fact the vortex model provides an account for the Universe on the basis of a single element. In addition to the accepted model of the quantum wave the model of the quantum vortex propositions that:

- *Mass is the quantity of vortex energy*
- *Static inertia is caused by the spin of energy in a spherical quantum vortex*
- *Kinetic inertia is set up by the propagation of energy in a transitional wave*
- *Potential energy is vortex energy*
- *The three-dimensional extensions of matter and space, electric charge, magnetism and gravity arise from the three-dimensional extension of the spherical vortex of energy*
- *The forces of electric charge, magnetism and gravity are caused by the interactions between quantum vortices of energy*
- *Infinite extensions of force fields and space are the infinite extensions of vortex energy*
- *Electric charge originates in the expansion or contraction of spheres of vortex energy*
- *Magnetism is set up by the rotation of concentric spheres of vortex energy*

- *Kinetics is caused by interactions between waves of energy and vortices of energy*
- *Wave particle duality stems from the bound state of vortices and waves of energy*
- *Space is the infinite extension of the vortex of energy beyond our direct perception*
- *Space curvature is caused by the spheres of space extending from stars and planets*
- *Electrons are light vortices with contracting concentric spheres of vortex energy*
- *Protons are massive vortices with expanding concentric spheres of vortex energy*
- *Neutrons are bound states of electrons and protons*
- *Mesons are captured energy swirling in proton and neutron quantum vortices*
- *Nuclear binding is caused by the swirl of captured energy between nuclear vortices*
- *Nuclear energy is the release of captured energy when nuclear vortices converge*
- *Sort lived particles are unstable synthesised vortices formed when energy is forced through natural stable quantum vortices*
- *Strangeness is a longevity conferred on an unstable synthetic vortex of energy by a stable natural quantum vortex at its core*
- *Particles of matter and antimatter are subatomic vortices with equal mass but opposite direction of vortex motion*
- *Gravity is a vortex interaction between matter and antimatter through the smallest realms of space*
- *Dark Energy is a consequence of vortex interactions between matter and antimatter through the largest realms of space*
- *The Fourth Dimension of Space influencing time is the dimension of size of space*
- *Time, set up by relationships between predictable events is space, is relative to the size of space, acceleration and the direction of the flow of energy in waves and vortices of energy.*

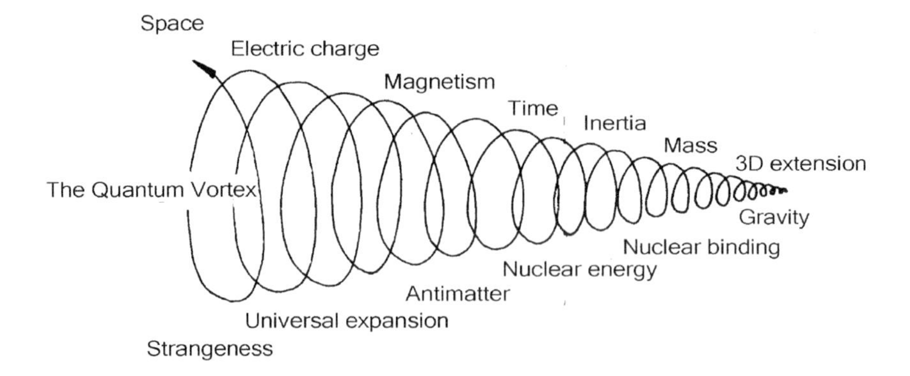

One day in India a group of four blind men came upon an elephant. One felt its leg and said it was a tree trunk. Another felt its ear and said it was a leaf. A third felt the elephant's trunk and insisted it was a python, while the fourth had gripped its tail and said it was a snake. Blind as we have been to the vortex of energy, while it is one thing we described it as many different things.

CHAPTER 12
QUANTUM FRAUD

In the subatomic vortex theory, matter is considered to be made up of just two types of stable vortex. One is the low mass electron vortex, which is negative in charge the other is the high mass proton vortex, which is positive in charge. The pi-meson, *pion*, is treated as energy captured in the proton. The neutron, which has no apparent charge, is considered to be a bound state of an electron and a proton.

Evidence for the neutron being an electron bound to a proton has been rejected in physics. It has been ignored, covered up or explained away. In science dismissing evidence is tantamount to fraud. I believe efforts have been made to disregard evidence concerning the structure of the neutron as it highlights a problem that arose in quantum physics which threatened it foundations.

Evidence for the bound state structure of the neutron includes the formation of neutrons when electrons and protons are attracted together, in a process called *K-capture*. Outside the atomic nucleus and in radioactive decay neutrons decay into an electron and a proton, in a process called *beta decay* and the mass of a neutron is slightly greater than the sum mass of an electron and a proton.

Despite the weight of evidence physicists refuse to accept that a neutron is an electron bound to a proton. Instead they insist that when an electron and a proton come together, to form a neutron, the electron and proton lose their identity altogether and then regain it again when the unstable neutron falls apart.

Physicists have gone to great lengths in quark theory to support their speculations of particle transmutations. In quark theory a proton is supposedly made up of two up-quarks with ⅔ charge and one down-quark with ⅓ charge, which add up to unitary charge. The neutron is supposedly made up of two down-quarks and one up-quark adding up to zero charge. In a stroke physicists used quark theory to explain away the bound state theory for the neutron. To shore up the theory they heaped more speculations into it.

According to quark theory when a proton interacts with an electron a *weak nuclear force* comes into play, which transforms an up-quark into a down-quark and the electron into an anti-neutrino that is temporarily lost. In this process the electron and proton cease to exist and their place is taken by the neutron and a neutrino. When a neutron decays, the proton is said to reappear with the anti-neutrino which then allows the electron to return and the neutrino to vanish and take away some energy with it as it goes.

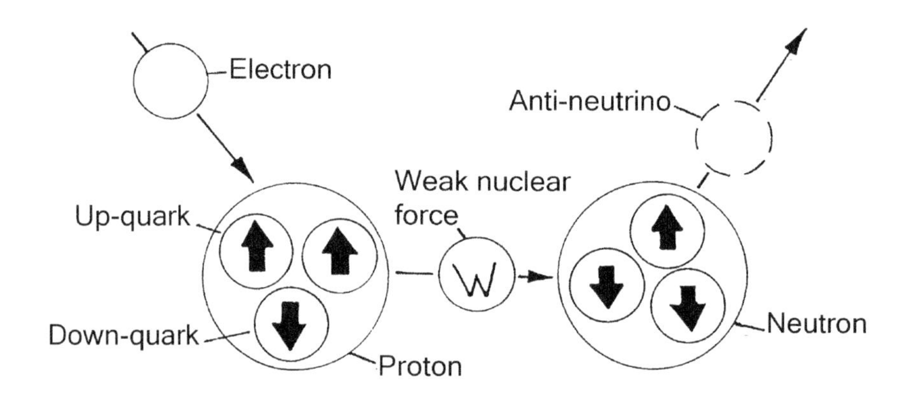

The view of the neutron, as a chargeless entity, rather than a bound state of two charges cancelling each other out, was challenged in 1957[1] by an experiment which revealed a very weak *electric dipole* on a neutron. On one spot the neutron displayed a minute negative charge - around a billion, trillion times weaker than that of a single electron. This appeared to support the idea that the neutron is a bound state of opposite charges, which mostly cancel each other out. That occurs in atoms which are electrically neutral because they contain equal numbers of oppositely charged particles.

Physicists were quick to dismiss the experiment on the grounds that the measure of electric charge in the neutron is too weak to be conclusive. In *Nuclei and Particles*,[2] Emilio Segré pointed out that the electric dipole moment of the neutron – equal to the charge on an electron x10^{-20} - is so minute compared to the margin of error - 0.1 +/- 2.4 - that *"this moment could be exactly zero, in agreement with the theory."*

1 Smith, J. H.; Purcell, E. M.; Ramsey, N. F. Experimental Limit to the Electric Dipole Moment of the Neutron, Physical Review, 108: 120–122, 1957

2 Segré Emilio, Nuclei & Particles (1964) Benjamin Inc

The electric dipole of a neutron would be expected to be very weak because if the neutron were an electron bound to a proton their opposite charges would all but cancel each other out but the measure should not be treated as exactly zero. Though the margin of error indicated the measure was too weak to be conclusive it did not suggest there was no measure at all. The way Segré wrote about the result of that experiment suggested to me that it revealed an awkward fact that needed to be hidden to protect 'the theory'. In my mind his wording confirmed my suspicion that the bound state model for the neutron posed a significant threat to quantum physics. Meanwhile I was collecting more evidence for the bound state theory, evidence that could not be so easily dismissed.

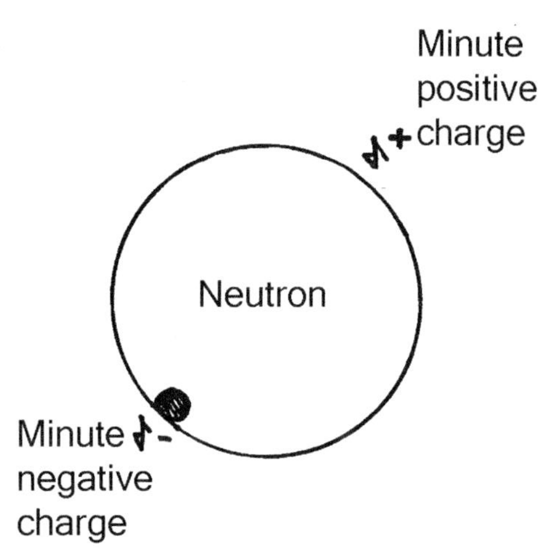

The neutron has magnetism. It has a magnetic moment of 1.91 nuclear magnetrons.[3] A particle cannot have a magnetic moment unless it has electric charge because the magnetic moment of a particle is created by the spin of its charge. Rather than accept the neutron must have a charge because it has magnetism, this evidence is ignored.

In high energy physics strange particles exit atomic nuclei with an electron at their centre. That suggests that there are electrons in atomic nuclei. If they are not in the nucleus bound in neutrons, where else in the nucleus could they be?

Chien Shiung Wu, a Chinese-American scientist, was a colleague of Segré. Known as the *First Lady of Physics*, she was a leading authority on neutrons. In 1956, Wu lined up the nuclei of radioactive atoms in a magnetic field so they were all spinning in the same direction, She observed, in beta decay, that more electrons were emitted in one direction than in another.[4] Wu's experiment revealed that when neutrons decay electrons are emitted directionally. This suggested they were sitting on a specific site on protons from which they emerged when the neutrons decayed. Wu's finding was more in line with the bound state theory than the quark theory. If electrons cease to exist when they come into contact with protons, to form chargeless neutrons, then reappear

3 Alvarez, L.W; Bloch, F. A quantitative determination of the neutron magnetic moment in absolute nuclear magnetrons Physical Review 57: 111–122. 1940.

4 . Richards et al, Modern University Physics, Part 2, Addison-Wesley 1973

when the neutrons decay, according to the quark model, their exit should be random in direction. There would be no reason why the electrons from beta decay should be directional. Also, Wu's neutrons could not have aligned in the magnetic field unless they contained charge because charged particles interact with magnetic fields whereas neutral particles do not.

Harald Fritzsch questioned the standard theory for neutrons having no electric charge on the grounds that if protons possessed energy for electromagnetic exchanges they should be more massive than neutrons whereas the evidence is they are less massive.[5]

Physicists argue that a neutron can't be an electron bound to a proton because a neutron has the same value of quantum spin as a proton or an electron. They say that if the neutron were an electron and a proton bound together it should display the quantum spin of them both. However, if a light electron were immobilised by a massive proton - nearly two thousand times as massive – it would not be free to spin. That is why it wouldn't possess quantum spin. (Quantum spin is the spin of a quantum vortex not the spin of energy in a quantum vortex.) Only the quantum spin of the proton would be apparent in a neutron if the electron is immobilised. The energy of quantum spin possessed by an electron, before it becomes locked onto a proton in the formation of a neutron, need not be lost. Captured as transient mass it could be suspended in the vortex complex of proton and electron. This could explain why the mass of the neutron is greater than the sum mass of the electron and proton. On its escape from a neutron, in beta-decay, the electron could carry away some of this captured energy, which would enable it to resume quantum spin as well as wave-propagation.

Physicists might argue that the law of *Conservation of Angular Momentum* does not allow for the quantum spin of an electron to be lost in a neutron. However, the neutron is unstable and physics does allow for conservation laws to be suspended in the formation and decay of unstable particles. In quantum mechanics so long as no conservation laws are broken in the overall process of formation and decay of unstable particles the conservation laws are satisfied to have been upheld.

I began to question the integrity of physics. Why were physicists so determined to dismiss evidence about the neutron and prop up the speculative quark theory instead. Before quark theory became generally accepted they were looking for evidence of quarks with fractional charge in experiments but found none until 1968 when, at the Stanford linear accelerator in California (SLAC), electrons were accelerated down the three-kilometer-long *vacuum*

5 Fritzsch H, Quarks: The Stuff of Matter, Allen Lane 1983.

tube by intense radio pulses and then targeted on protons in liquid hydrogen. The experiments showed that electrons were being scattered from what appeared to be something small and hard within the protons. Physicists were looking for quarks so interpreted the results of the experiments not only as evidence of quarks but of their fractional charge when in fact they didn't actually see quarks let alone fractional charges.

I was suspicious that the interpretation of the results of the experiments performed at the Stanford Linear Accelerator might have been influenced by a degree of desperation in physics to find an alternative to the bound state model for the neutron. Scientists live by the motto, *seeing is believing*, but I think, like most other people, they tend to see what they believe and overlook anything they don't believe in.

I objected to the quark account for neutrons as it flies in the face of the philosophy of science which is disinclined toward the heaping of speculations. Along with fractional charge, I was amazed at the blatant speculation of a *weak nuclear force,* which enabled the transformations of up-quarks into down-quarks. The inventions in quark theory were more worthy of alchemy than physics. The search for quarks reminded me of the desperate search in the Middle Ages for the Philosopher's Stone. Quarks are supposed to be bound by *gluon bonds* and physicists used this speculation to explain why they have never seen a quark. Medieval philosophers argued that if an angel were chopped in half two angels would emerge. This is because it would be impossible to have half an angel so two new angel halves would have to grow on each split end. Physicists say no one has ever seen a quark because when a quark is chopped out of a proton two new quarks immediately grow on the split ends of the gluon bond. This, they say, is because you can't have a gluon bond without a quark at either end. They contend that is why protons remain intact and two quarks mesons appear instead of individual quarks when protons are bombarded or collided in high energy experiments.

It is the story of the Emperor's New Clothing all over again. Quarks and other particles that have been invented in physics don't actually exist but few people in physics care to admit to it because there are so many jobs and academic reputations at stake. However, the advance of physics lies in the ability to face the awkward facts and question speculative theories. To quote George Gamow, *"Staggering contradictions of this kind, between theoretical expectations on the one side and observational facts or even common sense on the other are the main factors in the development of science."*[6]

I was encouraged when I read that quote and another by the inventor of

6 Gamow G., Thirty Years that Shook Physics, Heinemann.

quark theory, Murray Gell-Mann, "*Will some unknown young scientist find a new way of looking at fundamental physics that clarifies the picture and makes today's questions obsolete.*"[7]

I knew something was seriously wrong in physics and nothing short of a revolution would sort it out. The way that physicists argued so vehemently against the obvious structure of the neutron really alarmed me. It was one of the first objections I received when I introduced my embryonic theory to a physicist at Queens University in Belfast and decades later it blocked a possible presentation of my vortex physics to Prince Charles. Every time I introduced my vortex ideas to physicists they objected to the bound state model for the neutron which was essential in my theory. It is no wonder I was concerned at their determination that a neutron is not an electron bound to proton.

It was 1987 and I was working on my first book with Peter Hewitt when he alerted me as to why there was paranoia in physics at the idea a neutron is an electron bound on a proton. He drew my attention to the fact that it posed a deadly threat to *quantum mechanics.* (Quantum mechanics is not to be confused with the quantum theory established by Einstein. Einstein despised quantum mechanics.)

Quantum mechanics was founded on a *principle of uncertainty* published by Werner Heisenberg in 1927. He had proposed that the observer effect might apply at a quantum level. The *observer effect* is based on the idea that measurements of certain systems cannot be made without the act of observation affecting them. Heisenberg developed a principle to show that the process of making certain measurements in the sub-atomic world would increase the uncertainty about what was going on there. For example, if you wanted to look at a subatomic particle in order to ascertain its position you would reflect a quantum of electromagnetic radiation off it. But the act of bouncing a quantum of energy off a subatomic particle would give it a kick which would increase its momentum and that would make its position more uncertain. Looking at small objects requires more energy than is required for looking at large ones. This is evident in the electron microscope, which employs higher frequency radiation than an optic microscope. Because subatomic particles are the smallest things in nature the action of ascertaining their position with any degree of certainty would require the use of very large amounts of energy, which would give them an enormous kick. Heisenberg argued against physicists being able to determine, with any degree of certainty, both the position and momentum of a subatomic particle.

In addition to quantum indeterminacy other features of quantum theory,

7 Calder N. The Key to the Universe: A Report on the New Physics (1977) BBC Publications

including *wave-particle duality* determined by Louis de Broglie and *wave indeterminacy* proposed by Erwin Schrödinger, were incorporated to produce the *Heisenberg Uncertainty Principle.* This principle was central to the development of quantum mechanics from 1927 onwards. However, if Heisenberg's principle is applied to an electron locked onto a proton, somewhere in the circumference of a neutron, the high degree of certainty in its position would demands an enormous indeterminacy in its momentum.

When I was working with Peter Hewitt, on *The Vortex Key to Future Science*,[8] he was reading a book on quantum mechanics and drew my attention to a paragraph which said the neutron couldn't be an electron bound to a proton because the Heisenberg uncertainty principle predicted that electrons bound in neutrons could have velocities up to 99.97% of the velocity of light, and that was impossible. He showed me the paragraph and said it was proof my vortex theory for the neutron was wrong.

I retorted immediately, "No Peter, it is not my theory that is wrong, it is the uncertainty principle that is wrong."

Peter was silent. I was jubilant. In 1972 my father had sent a copy of my first sheet on the vortex theory to Sir Martin Ryle, then the Astronomer Royal. My father and Martin Ryle had known each other since they were boys. In his reply the Astronomer Royal said my theory was unsound because it suggested neutrons are electrons bound to protons and it also reminded him of medieval philosophers working out how many angels could fit on the head of a pin. He concluded his letter by saying that because quantum mechanics provided an adequate explanation for what is known in physics there was no place for my vortex theory. While his letter hurt the Astronomer Royal made me realise I would have to unseat quantum mechanics before my vortex theory would ever be taken seriously in physics.

There is no doubt that quantum mechanics has been an immense success in physics especially with the contributions by Max Born and Paul Dirac but its weakness was revealed when quantum physicists realised how the Heisenberg uncertainty principle allowed for untrammeled speculation. They speculated that forces could be carried between particles of matter by the exchange of short-lived particles, suggesting that within the bounds of the uncertainty principle, energy could be borrowed from the Universe to bring about the creation of short-lived *force-carrying particles*. So long as these force-carrying particles were sufficiently unstable to decay and repay the energy debt within the time allotted by Heisenberg's formula then no conservation law would have been broken in the overall process of their formation and decay.

8 Ash D & Hewitt P The Vortex: Key to Future Science, Gateway Books, 1990

If the time span of their existence was short enough the uncertainty principle allowed for very large amounts of energy to be involved in the formation of the particles, which in turn would enable them to carry very powerful forces.

In 1934 a young Japanese physicist, Hideki Yukawa, used the speculation of force carrying particles to predict the exchange of a meson force carrying particle, between protons, as a carrier of the strong nuclear force.

When, in 1947, *pi-mesons* were knocked out of atomic nuclei, in the cosmic

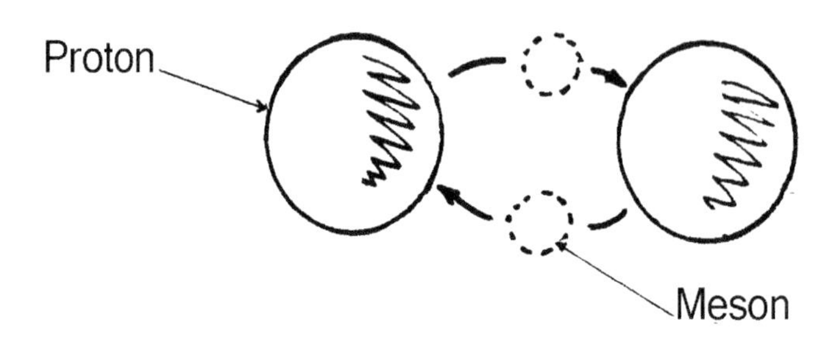

ray research carried out by Cecil Powell of Bristol University, Nobel Prizes followed for Yukawa in 1949 and Powell in 1950 and physicists throughout the world accepted the idea that forces are carried by the exchange of short-lived particles.

The underlying assumption needed for the new theory for forces was that 'anything is possible behind the screen of uncertainty'. The idea that sub-atomic particles can run lines of credit with the Universe, as though it were a bank, is ludicrous but because it is impossible to be certain that it doesn't happen no one could argue with claims that it does. Quantum physicists used uncertainty in the microcosm to argue the thesis for force carrying particles much as medieval philosophers used uncertainty in the macrocosm to argue how many angels could stand on a pinhead.

The fable for forces confirms a statement in the Talmud, which says: *'We see the world, not the way it is, but the way we are'.* Like most other people in the modern world, with a mortgage, bank loan, and credit cards, the professional physicist runs his affairs on credit. This projection was clearly apparent in quantum mechanics.

Mr. Proton is imagined to borrow energy to create gluons for the binding of quarks, W and Z particles for the weak nuclear force, mesons for the strong nuclear force, virtual-photons for electric charge and magnetism and gravitons for gravity. Attempts have been made to unify the theory into GUT – the General Unified Theory - so that the proton makes only a single demand on universal credit. In quantum theory the story of borrowing is undeniable.

Mr Proton

Borrowing that is never seen appears in another story. *The Borrowers...own nothing, borrow everything, and never forget their most important rule: you must never, ever be seen by the human beans.*[9]

Contrast this with the theory of the quantum vortex. Mesons exist and operate the strong nuclear force but they do not require the borrowing of energy within the bounds of uncertainty. They exist as real forms of energy with transient mass, captured within the proton vortex. This simple and obvious account for the meson is part an account for the forces of nature based on the single idea that subatomic particles are vortices of energy. Again Ockham's razor can be invoked for priority over the theory for forces in quantum mechanics.

It was a tragedy for physics when the British vortex theory was ditched at the turn of the 20th century. Since then physicists throughout the world have blindly followed the Copenhagen initiative, which disparaged 19th century models such as the vortex as naïve realism, which censured physicists from applying the vortex idea to subatomic particles.

The gentlemen physicists of Cambridge played cricket by the rules. When the Danish physicist, Niels Bohr, used quantum theory to account for the newly discovered *spectral lines* the British physicists accepted that as their theory for the vortex atom couldn't explain spectral lines it had been bowled out. They conceded to the Copenhagen team which took its innings under the captaincy of Bohr. Heisenberg was in bat for eight years with his quantum mechanics. Then he was bowled out by James Chadwick in the Cambridge team. When Chadwick discovered the neutron in 1935, Heisenberg should have finished his innings. But the Europeans, unfamiliar with the rules of cricket, continued to bat despite the fact that the uncertain wicket of quantum mechanics has been knocked out by the cricket ball neutron.

9 Norton M., The Borrowers, Puffin Books

Discovered at the Cavendish Laboratory in Cambridge, the neutron offered an experimental test for the uncertainty principle. If the neutron was treated as an electron bound to a proton then in that situation both the position and momentum of the electron could be defined with certainty. The position of the electron would be somewhere within the circumference of a neutron and the momentum of the electron could not exceed that conferred on it by the energy locked up in the mass of the neutron exceeding the sum mass of an electron and proton.

The rest mass of an electron is 0.911 x 10^{-30} kg. The rest mass of a proton is 1672.62 x 10-30 kg. The rest mass of a neutron is 1674.92 x 10^{-30} kg; therefore, it has a mass of 1.389 x 10^{-30} kg in excess of the sum mass of an electron and proton. This is equal to 1.5x the rest mass of an electron. In any neutron an electron could not possess a momentum in excess of that allowed by the energy locked up in 1.389 x 10^{-30} kg of mass by $E=mc^2$. [5]

If the uncertainty principle is applied to an electron, as though it were attached to a neutron, the certainty in its position would predict an enormous indeterminacy in its momentum. The uncertainty principle predicts that electrons bound in a neutron could have any velocity ranging from zero to 99.97% of the velocity of light. At this high-end velocity electrons would possess so much energy that their mass, according to $E=mc^2$ would be in the order of forty times the mass of an electron at rest. If neutrons were bound states of electrons and protons and the uncertainty principle were valid then some electrons in neutrons would have velocities approaching zero while others would have velocities approaching the speed of light and most would have a velocity somewhere in the range between. This wide range of velocities would be reflected as a range of mass values for neutrons from that of a single electron and proton to that of a proton and forty electrons. If the uncertainty principle were valid, neutrons would have mass values ranging from 1673.5 x 10^{-30} kg to 1710 x 10^{-30} kg. Experimental physics, however, has revealed that all neutrons possess the precise mass of 1674.92 x 10^{-30} kg.

Physicists were faced with a dilemma. If they accepted that the uncertainty principle had been disproved they would have been forced to abandon their theories for forces. The only alternative was to construct a theory to show that neutrons were not electrons bound to protons. They chose the latter course because by the time the neutron was discovered quantum mechanics had already gained so much momentum that physicists were reluctant to accept its uncertainty. They decided to ignore the evidence stacking up that the neutron was an electron bound to a proton until they could find or develop a solution to the problem. This came on the 1960s with the quark theory which enabled physicists to propose that the electron and proton ceased to exist as separate particles when they came together. Allowing that the two charged particles underwent a transmutation into an entirely neutral particle, they eliminated the awkward fact that an electron was bound to a proton in the neutron.

The Schrödinger wave equation had ensured it was impossible to be certain of an electron's position in an atomic orbit, which enabled atomic orbits to be excluded from the uncertainty principle. Now quark theory enabled physicists to dispense with the nasty little neutron as well.

Facts can be tweaked to fit theories but as Richard Feynman said, *"If your theories and mathematics do not match up to the experiments then they are wrong."* [7]

The truth is that since the discovery of the neutron physicists have been in denial of the evidence it presents because it threatens the uncertainty principle. In his book *Unravelling the Mind of God*[10], Robert Matthews said that Bohr and Heisenberg didn't really know where they were going but they set up a band wagon in physics that swept aside anyone who dared oppose them. One such was Albert Einstein who despised the emphasis placed on uncertainty in quantum theory in Physics. He knew Heisenberg's principle was unsound. At a Solvay conference Einstein argued through the night with Heisenberg reducing the younger man to tears. Einstein went on to describe the uncertainty principle as, "*...a real witches calculus... most ingenious, and adequately protected by its great complexity against being proved wrong.*" [10]

Many people will protest that the uncertainty principle must be right because of the incredible success of quantum mechanics. However, the successful application of a principle does not prove its validity. A car can run perfectly well without a certificate of roadworthiness but that does not mean it is roadworthy. A theoretical principle can work even if it is fundamentally flawed. With the failure of the uncertainty principle and the uncertainty about quark theory maybe physicists should be honest and admit that the Standard Theory is running without a certificate of science-worthiness. Perhaps the time has come for it to be scrapped.

The problem for physicists is that the credibility of science hangs on quantum mechanics. Quantum mechanics is the most successful theory in the history of science and is the core of advanced physics teaching in universities throughout the world. If physicists were to admit the uncertainty principle had been disproved by the neutron they would have to admit there has been a 'cover up' of evidence because it threatened their most cherished theory. That could shatter confidence in physics and confidence in the integrity of science.

10 Matthews R, Unraveling the Mind of God, Virgin Books, 1992

Physics is the king of sciences. If physics were to fall into disrepute then the credibility of science would be undermined. For the professional scientist that is unthinkable. To quote from *The Problems of Physics* by Professor A.J. Leggett:[11] *"Quantum mechanics...has had a success which is almost impossible to exaggerate. It is the basis of just about everything we claim to understand in atomic and sub-atomic physics, most things in condensed-matter physics, and to an increasing extent much of cosmology. For the majority of practicing physicists today it is the correct description of nature, and they find it difficult to conceive that any current or future problem of physics will be solved in other than quantum mechanical terms. Yet despite all the successes, there is a persistent and, to their colleagues, sometimes irritating minority who feel that as a complete theory of the Universe, quantum mechanics has feet of clay, indeed carries within it the seeds of its own destruction."*

The neutron is the particle that drives the chain reaction in nuclear weapons. It would be divine retribution if the detonator of the weapons of mass destruction developed by physicist, turned out to be their nemesis.

In his Inaugural Lecture entitled *'Is the end in sight for theoretical physics?'* [12] Professor Stephen Hawking declared that *"...because of the Heisenberg uncertainty principle, the electron cannot be at rest in the nucleus of an atom"* Electrons are at rest in atomic nuclei. That is why the end is in sight for theoretical physics.

When the neutron appeared as a black swan over the quaint little pond of quantum mechanics instead of welcoming its arrival, as scientific integrity would demand, physicists tried to shoot it. Had they let it land their cherished uncertainty principle would have been a dead duck and the virtual force-carrying particles, their darling little ducklings, would certainly have drowned. When they failed to shoot it they muddied the waters with incomprehensible math, white washed it with quark and with a flap about quantum spin chased it away.

A positive outcome of revealing uncertainty in the uncertainty principle is that it has led to uncertainty in quantum mechanics which in turn leads to uncertainty in the scientific criteria for truth. This is not a bad thing. Highlighting the importance of uncertainty Heisenberg may have released us from the burden of truth imposed by science. Religions have a history of telling people what they must and must not believe and increasingly there is a tendency for scientists to impose on us their understanding of the Universe with an equally

11 Leggett A.J. The Problems of Physics, Oxford University Press, 1987

12 Hawking S. Black Holes and Baby Universes, Bantam, 1993

obsessive degree of certainty. Heisenberg was a great and brilliant scientist. His ideas dominated physics throughout the 20th century. His lasting legacy may be his contribution to the infinity of possibilities by helping in the emancipation of humanity from the burden of religious and scientific certainty so that every human spirit can be at liberty to choose whatever they want to believe. By establishing the importance of uncertainty, Heisenberg may have done more for the truth than any saint or scientist because the paradox is that the less certain we are of truth the closer we are to it.

CHAPTER 13
THE NEUTRON

A question remains. If a neutron is an electron tightly bound to a proton why are neutrons unstable? In the free state neutrons have a half-life of eleven minutes. This means that in a group of neutrons after eleven minutes half of them will have decayed into an electron and proton and in the next eleven minutes their number would drop to a quarter etc.

Electrons are strongly attracted to protons by their opposite charge so why do they fall apart so easily? The answer may lie in the energy held by an electron.

The mass of the neutron, exceeding that of an electron and a proton by the mass of one and a half electrons, could be the mass equivalent, by $E=mc^2$, of the kinetic energy an electron holds, which keeps it flying around. When the electron is locked onto a proton this kinetic energy would resist the electro-static attraction between the electron and proton that binds them together in the neutron.

To understand this situation, imagine a ping-pong ball bouncing in a water fountain. It is only a matter of time before the ball bounces away from the fountain. In this analogy the ping-pong ball represents the electron attached to the surface of a proton to form a neutron and the force of water in the fountain represents it's kinetic energy. It is only a matter of time before the kinetic energy of the electron would succeed in overcoming its electro-static attraction to the proton, enabling it to break free. That could explain the decay of the neutron. This would cause the electron to bounce on the proton.

Now imagine you are watching a group of fountains. Each has a ping-pong ball bouncing on it. Every now and then a ball bounces free of the fountain. The escapes are random but if you record the time it takes for half the balls to escape their fountains and then how long it takes for half the remaining balls to bounce free you would notice that the time interval between the numbers of balls left on the fountains between a half and a quarter is the same as between the whole and the half. This represents the half-life of table-tennis balls bouncing on fountains, which would enable you to calculate the probability of a ball escaping a fountain. This analogy can help to understand half-lives in radioactive decay.

In the nucleus of an atom the half-life of radioactive decay differs from element to element. In the nuclei of most atoms in the periodic table of the elements the neutrons appear to be stable and there is no radioactive decay. In the study of beta decay it is apparent that electrons are more likely to escape atomic nuclei if they are exceptionally large, or very small. The reason I give for this anomaly is that the apparent stability of neutrons in the nuclei of most atoms could be due to electrons being caught by other protons in the nucleus when they break away from the neutrons therein. The chance of an electron escaping would be greater from a large loosely knit group of nucleons than a smaller tightly knit group.

My belief is that neutrons are not stable in atomic nuclei. I believe electrons are constantly being exchanged between protons and neutrons in the nucleus. I think it is a very dynamic situation. The reason, I believe, why most nuclei are not subject to beta decay is that if neutrons are protons with an electron while protons are without electrons then there will always be more protons than electrons in a nucleus, which means as soon as an electron breaks free of one proton it is likely to be caught by another. I reckon the predominance of protons over electrons makes it unlikely for an electron to escape from an atomic nucleus but if electrons do escape it would be more likely from a large or small group of protons and neutrons. It also appears to depend on the ratio of protons and neutrons in an atomic nucleus. I use an analogy of cats and mice to illustrate this.

I liken electrons in the nucleus of an atom to mice and the protons to cats and the tails of the mice to the kinetic energy of the electron held in partial capture. Apart from the simple hydrogen atom formed when an electron orbits a single proton, most atomic nuclei are a mix of protons and neutrons. I imagine them as a group of cats in which some cats have a mouse in their jaws or between their claws and some have none. As soon as a mouse breaks away from a cat with a mouse it is caught by another cat which doesn't have a mouse. So long as there is a predominance of cats over mice it is unlikely a mouse would break free of the group.

If a cat is on its own with a very lively mouse, the mouse would have a better chance of escape than if it were in a group of cats continually pouncing on it as fast as it attempted an escape. This is why neutrons in a nucleus are less evident in their decay than neutrons in a free state.

If you think about the mouse analogy a mouse would be more likely to break out of a larger, looser group of cats and mice than a medium size more coherent group. If the group is small with overall fewer cats the mice might have a better chance to escape than from the medium size group because of their higher surface area to volume ratio. But in exceptionally large groups, which are a lot less tightly packed and so have more space between the cats, there would be a better chance for a determined mouse, or even cats with mice, to break free.

I use the cat and mouse analogy to explain radioactivity. Atoms, like uranium, with large, loosely bound nuclei, tend to be radioactive. They are represented by large groups of cats in my analogy. Radioactivity comes in the forms of high energy gamma ray photons, beta decay electrons represented by mice in the analogy and alpha-decay helium nuclei represented by four cats escaping with two mice continually tossed between them. The loss of neutrons from a nucleus is represented by a stray cat with a mouse in its jaws.

Escaping mice pay a price. As a mouse breaks free of a group of cats it appears to lose a bit of tail to tooth or claw. That was how I use the cat and mouse analogy to account for the fact that electrons come away from radioactive atoms with a wide range of energies in beta decay. I use capture theory to account for this. When an electron vortex is locked on the surface of a proton vortex, in the formation of a neutron, its kinetic energy would be held in complete capture by the proton. But because the proton is already saturated with energy in complete capture this could set up a *supersaturated state of capture* that would be innately unstable. This instability, setting up a tussle between mass and kinetic energy, exemplified in the ping-pong analogy, would enable electrons to break free of the proton and account for the decay of the neutron. In the process it could be that the amount of energy that comes away with the electron as kinetic energy varies because some of the energy it held in partial capture could be ripped away and remain in a state of supersaturated capture by the proton. I used the analogy of part of the tail of a mouse being ripped away by a cat as it escapes tooth of claw to illustrate this and provide an account for *neutrinos.* Neutrinos were particles invented by Wolfgang Pauli, in 1930. They were supposed to carry away energies that were otherwise unaccounted for in beta decay.

I also used the cat and mouse analogy to explain the detonation of nuclear weapons. To begin with it is necessary to understand the concept of *isotopes.* In the same element atoms can vary in the number of neutrons they contain. These are called isotopes. Uranium (U), for example, has three naturally occurring isotopes, U^{234}, U^{235} and U^{238}. Each isotope of uranium has the same number of 92 protons but an atom of Uranium^{234} has 142 neutrons, an atom of uranium^{235} has 143 neutrons and an atom of uranium^{238} has 146 neutrons. Uranium^{238} is the most common isotope of uranium. Uranium^{235} is less plentiful and Uranium^{234} is the rarest natural isotope being the breakdown product of Uranium^{235} after it has shed a neutron. Uranium^{235} (U^{235}) is fissile. If a neutron enters its nucleus the nucleus will split in two and release two to three more neutrons. In my cat and mouse analogy I depict U^{235} as having a metaphorical cat-flap that lets in stray cats looking for a home. However, when a stray with its mouse, enters the uranium^{235} home it so disturbs the equilibrium of the cats already there that they fall out. The feline congregation first becomes turbulent, then it swells and finally it splits in two, releasing pent-up energy from the group. This represents nuclear energy. At the same time two or three cats, each with a mouse in its jaws, get kicked out. These jawful strays then look for new homes and if they find more of the uranium^{235} atoms, with the allegorical cat-flaps, they enter in uninvited and split the resident groups releasing more furious felines, fire and spitting fury.

Physicists in the Manhattan Project isolated, purified and concentrated uranium^{235} and then crushed it into a 15kg ball. As they did so a chain reaction began. Neutrons, released from the natural decay of U^{235} to U^{234}, entered U^{235} nuclei and split them to release two or three neutrons. Each one then went on to split two or three more U^{235} nuclei causing the release of four or six more neutrons which then split another four to six U^{235} nuclei to release eight to twelve neutrons then sixteen to twenty-four, thirty-two to forty-eight and so forth. This chain reaction rapidly broke up ever more nuclear households. With the split of each U^{235} nucleus into smaller, tighter nuclei, nuclear energy was released and in an exponential flash, as the number of splitting nuclei grew to countless trillions, a nuclear explosion occurred.

CONCLUSION

Science has become the religion of the modern age. It is a cult following the figure of Democritus. The cult of Democritism is based on a doctrine derived from his false philosophy of materialism. It has strong parallels with medieval Catholicism except that science is concerned only with the seen whereas Catholicism was preoccupied with the unseen. The pride of science is impartiality. There is no impartiality in Democritism and woe betide anyone who questions its dogma of physicality. The cult of Democritism is destroying science and the soul and spirit of humanity. It is robbing us of our freedom of self-determination. The high priests of this religion are living a lie. Burying the evidence pertaining to natural particles they have invented nebulous hypothetical particles instead. They are in a constant chase to find the ultimate atoms predicted by Democritus but the endeavours are futile because, as Einstein pointed out over a century ago, the ultimate particles are quanta of non-substantial energy. The substantial particles envisaged by Democritus do not exist. There exist only particulate forms of energy which seem to be more in the nature of thoughts than things. Increasingly quantum physicists view the Universe as a mind whereas in Democritism it is viewed as something more mechanical.

The invention of quarks as ultimate particles of matter was a prime example of the quackery in Democritism. After decades of searching for quarks physicists are confident they have been detected but the detection of quarks has been more an interpretation of experimental data than the discovery of an actual particle.

The invention of nebulous particles is more to do with faith than physics; faith in the atomic hypothesis of Democritus that at a fundamental level particles can explain everything. That has led to an epidemic of particle speculations in physics. This issue was addressed on September 26, 2022 in the *Guardian.*

In an article headlined: ***No one in physics dares say so, but the race to invent new particles is pointless,*** the astrophysicist, Sabine Hossesnfelder, laid bare the truth about the pointless pursuit of particles in physics. The vortex account for how energy forms mass is much needed in physics so we can shed particle materialism and better understand quantum theory and the world we live in.

Kelvin was on track when he embraced the vortex model for the atom proposed by Helmholtz. Kelvin said *"the material atom is a monstrous assumption..."* as it did nothing to define the fundamental properties of matter. Had the vortex initiative survived the 20th century I am sure it would have been applied to energy instead of the ether and to subatomic particles instead of atoms. Unfortunately, in the quantum revolution many useful models from 19th century physics, including the vortex, were discarded as naïve realism, which left the way open for the deeply entrenched naïve realism of particle materialism to grow in strength throughout the 20th century and on into the 21st.

People need something to believe in. In the past it was religion. Today it is science. But science has not done away with religion, it has become the new religion instead.

Reading Michael Dine's book, *This Way to the Universe: A Journey into Physics,*[1] it was clear that physicists, obsessed with the falsehoods in Democritism, have increased the distance between their understanding of physics and people outside the physics community. The subject has become so complex it is now far less understandable than ever it was when I was a student. This is a failing in physics because, as Heisenberg wrote, *Even for the physicist the description in plain language will be a criterion of the degree of understanding that has been reached.*[2]

In the history of science descriptions in plain language were valued alongside math. This ensured a degree of understanding could be reached before the mathematics became so complex it took the understanding out of reach.

I believe that in physics today too much emphasis has been placed on mathematic. As Gary Zukav wrote in his book *The Dancing Wu Li Masters,*[3] *The fact is that physics is not mathematics...Stripped of mathematics, physics becomes pure enchantment.*

Physics is awash with mathematical formulae but there is a lack of fundamental understandings. This was made clear by Richard Feynman when he said,

1 Dine M. This Way to the Universe: A Journey into Physics, Penguin Books, 2023

2 Heisenberg W. Physics and Philosophy Harper & Row, 1958

3 Zukav G., The Dancing Wu Li Masters, Rider, 1979

"...we have the formulas for that, but we do not have the fundamental laws."

In the footsteps of my father I am an amateur in the field of physics, which is fortunate as it enabled me to maintain simplicity. He was an advocate of Lord Ernest Rutherford and encouraged me to follow Rutherford's maxim, *"These fundamental things have got to be simple."*

My father also taught me not to seek recognition. As Max Planck famously remarked, *"Science advances from funeral to funeral."* He also said, *"A scientific truth does not triumph by convincing its opponents and making them see the light, but because its opponents eventually die and a new generation grows up that is familiar with it."*

The quantum vortex theory started out as a line of enquiry to see if the theory of the vortex atom might apply to subatomic particles. The outcome was radical, perhaps too radical for the worldwide community of physicists to accept, embedded as they are to materialism and opposed as they are to any non-physical directions in science. Expectations of recognition in this generation are unrealistic. My hope is that this work will reach into the ground roots to benefit future generations who might be more open to spiritual directions than the present generation of scientists.

The quantum vortex theory is not complete as a physics. It is a route map of possibilities offering directions for anyone who might be interested in the concept of the subatomic vortex of energy and the implications not only in the physical world but in worlds beyond the physical. The theories of the quantum vortex and superenergy are intended to offer a fresh approach to science and religion in the hope that a new platform of understanding will emerge which will embrace the best elements of both to provide a better understanding of the Universe and our place within it. Quantum physics has the potential to achieve this but not quantum mechanics.

The ignorance of evidence in regard to the neutron, to protect the postulate of uncertainty in quantum mechanics, is an outstanding example of the ignorance that is rife in science. Evidence of supernormal phenomena is routinely ignored. The practice of disregarding evidence should be eradicated along with the bias toward materialism in science, which is at the root of this practice. Practically every university in the world is complicit in covering up evidence that does not comply with the doctrine of physicalism. Ignoring evidence is anathema in science. Ignoring evidence, especially in regard to the neutron undermines its integrity.

The theory of the quantum vortex can help us step clear of materialism and see beyond physicalism to discover purpose through a deeper understanding of the Universe. A scientific theory that offers meaning to existence has to be better than a material hypothesis that concludes the Universe is purposeless, a "...*meaningless hodgepodge of atoms.*"

The materialistic approach in science was necessary to rescue mankind from religious atrocities and superstition. But it has gone too far. In the name of skepticism, society is tending to follow scientific materialism without question. This is the very antithesis of skepticism which is to question everything. Directions of research are cherry picked in favour of the material ethos and anything remotely spiritual is ignored, discredited or blocked from serious investigation.

However, the theory of superenergy is not so easy to ignore because it is inextricably linked to the quantum vortex theory which threatens to blow science apart. The theory has revealed dishonest practices in physics, which undermine the integrity of science. The disregard of evidence pertaining to the neutron in order to protect quantum mechanics crashes the credibility of science. Before anyone takes to picking the theory of the quantum vortex to pieces they must first answer the challenge it presents to the cornerstone of quantum mechanics; the uncertainty principle of Werner Heisenberg.

There is a lot at stake here. Governments follows scientific opinion. Education is based on the benchmarks of science and the media are biased in its favour. The trustworthiness of science underpins the whole of society and the moral fabric of humanity.

The bias in science against spirituality and supernatural phenomena is based on prejudice, not evidence. The evidence coming from a tsunami of near death experiences points to a continuity of life after death. This is a matter of great concern to us all yet news of it rarely reaches the media and scientific institutions, for the most part, prefer to ignore it.

The problem with scientists and scientific institutions is that they cannot be trusted. Anything that opposes consensus beliefs in science is deemed unscientific. The opposition to spirituality is not evidence based. It is ideological. When scientists say there is a lack of evidence for spiritual phenomena it is largely because they have ignored the evidence and blocked research into supernormal subjects.

The theory of superenergy represents an attempt to provide a theoretical framework of understanding to help in the comprehension of spiritual and supernatural phenomena. The superenergy theory hinges on the theory of the quantum vortex which stands in its own right as a sound scientific theory. I present these theories as seeds for the field of knowledge but before they can be planted the ground must first be cleared of the weeds that are choking both religion and science.

ILLUSTRATIONS

INDEX

B

C

D

K

L

M

N

O

P

Q

R

S

David A Ash

ABOUT THE AUTHOR

David Ash was born in Kent, England in 1948. His father, Dr Michael Ash, researching the link between cancer and radioactivity, introduced him to radioactivity physics when he was seven.

He discovered the vortex theory for the atom in 1965, during his last year at school, and began to collect ideas and data on subatomic matter in 1967 at the Hastings college of further education. David started work on the vortex theory at Queens University of Belfast in 1968, after Professor Gareth Owen, Dean of the Faculty of Science, encouraged him to use his time at university to develop a theory rather than cram for a career.

David went on to read physics and work on the vortex theory at Queen Elizabeth College, London University where he graduated with a bachelors science degree in 1972. After a gap year he went onto the College of St Mark and St John in Plymouth in 1973 where he secured a postgrad in physics education in 1974.

Whilst there he was introduced to the Royal Institution and joined it as a young member. On January 15th 1975 he gave a lecture on the quantum vortex to young members and guests, from the rostrum where Lord Kelvin had demonstrated his vortex theory for the atom a century before.

David left London in 1976 to join his father who had founded CAMS, the College of Alternative Medicine and Science. There he taught the vortex theory, which was then published in part (with Peter Hewitt) in *The Vortex: Key to Future Science*, by Gateway books in 1990. After that David set out on an international lecture tour between 1991 and 1994.

In 1995 the College of Psychic Studies published his vortex theory in full as *The New Science of the Spirit*, which included a prediction of the accelerating expansion of the Universe. An updated version of the vortex theory was published in 2001 as *The New Physics of Consciousness* by Kima Global Publishing of Cape Town, followed by *The Vortex Theory* as a further update in 2015. He included the vortex theory in AWAKEN published by Kima in 2018. Written with the help of Matthew Newsome and David's partner Susan Thompson, AWAKEN was revised in 2021.

David is father to nine and has fifteen grandchildren.

Mill House Publishers s/e
Roldvej 10, 5492 Vissenbjerg,
Assens, Denmark
https://www.millhouse-publishers.com

Printed in Great Britain
by Amazon